Topics in Safety, Risk, Reliability and Quality

Volume 28

More information about this series at http://www.springer.com/series/6653

Kevin MacG. Adams

Non-functional Requirements in Systems Analysis and Design

Springer

Kevin MacG. Adams
Information Technology and Computer Science
University of Maryland University College
Adelphi, MD
USA

ISSN 1566-0443 ISSN 2215-0285 (electronic)
Topics in Safety, Risk, Reliability and Quality
ISBN 978-3-319-18343-5 ISBN 978-3-319-18344-2 (eBook)
DOI 10.1007/978-3-319-18344-2

Library of Congress Control Number: 2015937211

Springer Cham Heidelberg New York Dordrecht London

Printed on acid-free paper

Springer International Publishing AG Switzerland is part of Springer Science+Business Media (www.springer.com)

To Admiral Hyman G. Rickover, Vice Admiral Levering Smith and Rear Admiral Wayne Meyer who conquered bureaucratic engineering through technical excellence and personal perseverance to produce complex systems that have ensured our liberty through strength and vigilance.

Kevin MacG. Adams
University of Maryland
University College

Preface

At the end of the last century, corporations and government entities in the United States showed increasing concern for the loss of competitive advantage previously enjoyed by products designed and manufactured in the United States. The loss of competitive advantage experienced by manufacturers of these products was attributed to a variety of causes that both threatened the country's standard of living as well as its position within the larger world economy (Dertouzos et al. 1989). A report by the National Research Council reported that:

> *Engineering design is a crucial component of the industrial product realization process. It is estimated that 70 percent or more of the life cycle cost of a product is determined during design.* (NRC 1991, p. 1)

The engineering community agreed with this assessment, stating that "market loss by U.S. companies is due to design deficiencies more than manufacturing deficiencies" (Dixon and Duffey 1990, p. 13). A variety of studies on both manufacturing and engineering design were undertaken in order to improve the situation in both the industrial sector and in academia (NRC 1985, 1986, 1991). The engineering community found that in order to improve both the cost and efficacy of products produced for the global economy and "To regain world manufacturing leadership, we need to take a more strategic approach by also improving our engineering design practices" (Dixon and Duffey 1990, p. 9).

"Market loss by U.S. companies is due to design deficiencies more than manufacturing deficiencies" (Dixon and Duffey 1990, p. 13). The engineering design process in use in the industrial sector required improvement, but more importantly, the theory of design and implementing design methodologies advocated by the academic community were stagnant. A renewed emphasis on design and a new subdiscipline in engineering design were adopted by the engineering community. New requirements for design activities in the academic curricula were mandated, and the national engineering accreditation organization included additional design criteria as part of the accreditation assessment process. Major efforts to reinvigorate design in both undergraduate and graduate engineering programs in the United States have reemphasized the role of design in the engineering curricula. This text has been developed to address a unique topic in engineering design, thereby filling a void in the existing engineering literature.

The topic of this text—*Nonfunctional Requirements in Systems Analysis and Design*—supports endeavors in the engineering of systems. To date, nonfunctional requirements have only been addressed within highly focused subdisciplines of engineering (e.g., reliability, maintainability, availability; traceability; testability; survivability; etc.). The wider engineering community has not had access to materials that permit them to develop a holistic, systemic perspective for nonfunctional requirements that regularly affect the entire system. Having a basic understanding of how the principal nonfunctional requirements affect the sustainment, design, adaptability, and viability concerns of a system at a high level, should help fill a void in the existing engineering literature.

To support this approach to understanding nonfunctional requirements during engineering design endeavors, the book is divided into six major parts: (1) Systems Design and Nonfunctional Requirements; (2) Sustainment Concerns; (3) Design Concerns; (4) Adaptation Concerns; (5) Viability Concerns; and (6) Conclusion.

Part I focuses on the purposeful design of systems and how nonfunctional requirements fit into the design approach. Chapter 1 provides an introduction to the design of engineering systems, and reviews how engineers in the various engineering disciplines are responsible for developing designs for complex man-made systems. It also addresses systematic design, the breadth and depth of disciplines associated with design activities, and the use of life cycle models and supporting processes in the systems design process. Chapter 2 provides a description of engineering design and explains how it fits within the larger scientific paradigm. It includes a description of the desirable features and thought processes invoked in good engineering design methodologies. The chapter contains a high-level overview of seven historically significant design methodologies. It concludes with a more detailed section on axiomatic design and explains why axiomatic design is proposed as an effective system-based approach to the design of engineering systems. Chapter 3 provides a formal definition for nonfunctional requirements and the role they play in the engineering design of man-made, complex systems. It addresses the wide range of nonfunctional requirements, and introduces a number of taxonomies that have been used to describe nonfunctional requirements. The chapter concludes by presenting a notional taxonomy or framework for understanding nonfunctional requirements and their role as part of any system design endeavor. This taxonomy distributes 27 nonfunctional requirements into four *concerns*: sustainment concerns, design concerns, adaptation concerns, and viability concerns. The four concerns serve as the headings for the next four Parts of the text.

Part II addresses sustainment concerns during systems design endeavors. It is divided into two chapters which address five nonfunctional requirements. Chapter 4 addresses the nonfunctional requirements for reliability and maintainability. The section on reliability reviews the basic theory, equations, and concepts that underlie its utilization, addresses how reliability is applied in engineering design, and also explains how reliability is used as a technique for determining component reliability. The section concludes with a metric and measureable characteristic for reliability. The section on maintainability defines basic terminology, how maintainability is used in engineering design, and introduces the maintenance and support concept. It concludes with a metric and measureable characteristic for maintainability. Chapter 5 addresses the nonfunctional requirements of availability, operability, and testability. The topic on availability and operability introduces basic theory, equations and concepts that underlie availability, how availability is applied in engineering design, and concludes with a metric and measureable characteristic for reliability. The second major topic in the chapter defines testability, discusses how it is used in engineering design, establishes a relationship to availability, and concludes with a metric and measureable characteristic for testability.

Part III addresses design concerns during systems design endeavors. It is divided into three chapters which address nine nonfunctional requirements. Chapter 6 addresses the nonfunctional requirements for conciseness, modularity, simplicity, and traceability. The topic on conciseness reviews the basic terminology, equations, and concepts that underlie its utilization, and proposes a metric for measuring and evaluating conciseness. The next section discusses the concept of modularity and how it affects systems designs. A number of specific modularity measures from the extant literature are presented. The section concludes with the selection of a measure for modularity, and presents a structural map relating the metric and the measurement attributes for modularity. The section on simplicity contrasts it with complexity, reviews relevant measures for complexity, and presents a measureable characteristic for complexity. The chapter concludes by discussing traceability, in relation to how it impacts system design endeavors, and develops a metric for evaluating traceability in systems designs. Chapter 7 addresses the nonfunctional requirements for compatibility, consistency, and interoperability. The chapter begins by reviewing compatibility and the basic terminology, equations, and concepts that underlie its utilization. Compatibility and its relation to standards is addressed, and a measure for evaluating compatibility in systems design is provided. The second section discusses the concept of consistency in terms of how it affects systems designs, and proposes a measure for consistency that is based upon requirements validation, functional verification, and design verification activities. The final section in this chapter addresses interoperability by providing both a definition and models of interoperability, and proposes a formal method for evaluating interoperability. Chapter 8 addresses the nonfunctional requirement for safety. The chapter contrasts *machine age* systems safety with *systems-age* concerns, and provides a system-based model for system safety. The chapter concludes by relating the proposed measure for evaluating systems safety to a metric, and includes a structural map for systems safety.

Part IV addresses adaptation concerns during systems design endeavors. It is divided into two chapters which address nine nonfunctional requirements. Chapter 9 addresses the nonfunctional requirements for adaptability, flexibility, modifiability and scalability, and robustness. The chapter begins with a section that reviews the concept of changeability, its three unique elements, and presents a method for representing systems change using a state-transition-diagram. Both adaptability and flexibility are defined, and a method for distinguishing between these two nonfunctional properties is proposed. Modifiability is defined, and a distinction between it and both scalability and maintainability is provided. Robustness is defined, and its impact on design considerations is discussed. The chapter concludes by defining a measure and a means for measuring changeability that is a function of all four nonfunctional requirements discussed in the chapter. Chapter 10 addresses the nonfunctional requirements for extensibility, portability, reusability, and self-descriptiveness. The chapter begins by reviewing extensibility, its definitions, and how it is approached as an aspect of purposeful systems design. Portability is defined, positioned as a desirable characteristic, and is discussed as it relates to the four factors designers must consider in order to achieve portable designs. Reusability is addressed by providing both a definition and an explanation of its role in systems designs. Both top-down or bottom-up approaches, and three unique techniques that address reusability are presented. The section concludes by recommending two strategies and ten heuristics that support reusability in systems designs. Self-descriptiveness is defined and discussed by emphasizing the types of problems associated with poor self-descriptiveness. Seven design principles for user-systems dialogue are proposed to decrease errors and improve system self-descriptiveness. The chapter concludes by defining a measure and a means for measuring adaptation concerns, which is a function of extensibility, portability, reusability, and self-descriptiveness.

Part V addresses viability concerns during systems design endeavors. It is divided into two chapters which address eight nonfunctional requirements. Chapter 11 addresses the nonfunctional requirements for understandability, usability, robustness, and survivability. The first three nonfunctional requirements are defined and positioned within the requirements for good system design. The fourth nonfunctional requirement, survivability, is defined and 17 design principles that may be invoked when designing for survivability are addressed. The chapter concludes by defining a measure and a means for measuring core viability concerns, which is a function of understandability, usability, robustness, and survivability. Chapter 12 addresses the nonfunctional requirements for accuracy, correctness, efficiency, and integrity. The chapter begins by reviewing accuracy, its definitions, and concepts related to reference value, precision, and trueness. The second section defines correctness, and demonstrate how both verification and validation activities provide evaluation opportunities to ensure correctness. Four design principles that support the development of systems that correctly represent the specified requirements for the system are reviewed. Efficiency is addressed by providing a clear definition for efficiency, and by establishing a proxy for system efficiency. Integrity and the concept that underlies its use as a nonfunctional requirement in systems designs is reviewed. Thirty-three security design principles, and the life cycle stages where they should be invoked when designing for systems for integrity are proposed. The chapter concludes by defining a measure and a means for measuring other viability concerns, which is a function of accuracy, correctness, efficiency, and integrity.

Part VI provides a conclusion in Chap. 13. The conclusion reviews the climate that led to the small crisis in engineering design during the late 1980s and the need for revision of the engineering curricula and accreditation criteria. The major efforts to reinvigorate design in both undergraduate and graduate engineering programs in the United States which reflected the reemphasis of the role of design in the engineering curricula are covered. Finally, the rationale for the development of the text, and the need to address nonfunctional requirements in systems analysis and design endeavors are reviewed.

This book is intended for use by systems practitioners or in a graduate course in either systems engineering or systems design where an understanding of nonfunctional requirements as an element of the design process must be understood. Given its discipline-agnostic nature, it is just as appropriate for use in a software, mechanical, or civil engineering class on design or requirements. The book may be utilized in a traditional 12- or 14-week schedule of classes. Part I should be taught in order of appearance in the book to provide the proper theoretical foundation. Parts II–V can be taught in any order, although, lacking any other preference, they can be taught in the order in which they appear. The conclusion in Chap. 13 should follow the conclusion of Parts I–V, as it builds on the information developed in Chaps. 4–12.

Upon completion of the text, the reader or student should have an improved understanding and appreciation for the nonfunctional requirements present in complex, man-made systems. Although the text addresses only 27 nonfunctional requirements, the author recognizes that many additional nonfunctional requirements exist and that they may be required to be addressed in many systems design endeavors. However, armed with the approach used in understanding the 27 defined functional requirements (i.e., definition, design utilization, measurement, and evaluation), additional nonfunctional requirements may be similarly treated.

As always, the author takes responsibility for the thoughts, ideas, and concepts presented in this text. Readers are encouraged to submit corrections and suggestions through correspondence with the author in the spirit of continuous improvement.

References

Dertouzos, M. L., Solow, R. M., & Lester, R. K. (1989). *Made in America: Regaining the Productive Edge*. Cambridge, MA: MIT Press.

Dixon, J. R., & Duffey, M. R. (1990). The neglect of engineering design. *California Management Review, 32*(2), 9–23.

NRC. (1985). *Engineering Education and Practice in the United States: Foundations of Our Techno-Economic Future*. Washington, DC: National Academies Press.

NRC. (1986). *Toward a New Era in U.S. Manufacturing: The Need for a National Vision*. Washington, DC: National Academies Press.

NRC. (1991). *Improving Engineering Design: Designing for Competitive Advantage*. Washington, DC: National Academy Press.

Acknowledgments

I would like to start by acknowledging three inspirational naval engineers that led the most successful naval engineering endeavors of the twentieth century. Their legacy has directly affected me and my perspective on engineering.

- Admiral Hyman G. Rickover [1900–1986], *Father of the Nuclear Navy*, led the group of engineers, scientists, and technicians that developed and maintained the United States Navy's nuclear propulsion program from its inception in 1946 until his retirement in 1983. Acknowledged as "the most famous and controversial admiral of his era," (Oliver 2014, p. 1) Admiral Rickover's personal leadership, attention to detail, conservative design philosophy, and program of intense supervision ensured that the *Nuclear Navy* has remained accident-free to this day. I was privileged to serve, in positions as an enlisted machinist's mate, a submarine warfare officer, and a submarine engineering duty officer in Admiral Rickover's program for over 23 years. Admiral Rickover's legacy was present in all we did. The magnificent underwater vessels that continue to protect this country are a tribute to both his engineering brilliance and his ability to persevere in the face of great odds.
- Vice Admiral Levering Smith [1910–1993] served as the first technical director for the Polaris submarine launched ballistic missile program, "the most convincing and effective of the nation's strategic deterrent weapon systems" (Hawkins 1994, p. 216). He served in this capacity from 1956 until his retirement in 1974. During this time, Vice Admiral Smith led the team that conceived and developed Polaris, transitioned the force from Polaris to Poseidon, and led the conceptual development of the current Trident ballistic missile system. The legacy of the Navy's strategic systems program "may have set an unattainable standard for any equally important national endeavor" (Hawkins 1994, p. 215). Once again, I was privileged to serve on a Polaris-Poseidon capable submarine for a period of five years. Vice Admiral Smith's technical acumen was present throughout the weapons department and the field activities providing support for our operations.

- Rear Admiral Wayne Meyer [1926–2009], *Father of Aegis,* guided the development and fielding of the Navy's Aegis weapons system, the *Shield of the Fleet*, from 1970 until 1983. Rear Admiral Meyer changed the way Navy surface combatants were designed, built, and delivered to the Navy. The first Aegis warship, "TICONDEROGA and her combat system were the product of a single Navy Project Office (PMS 400), led by RADM Wayne E. Meyer and assigned the total responsibility for designing, building, deploying, and maintaining the AEGIS fleet. The creation of this single office with total responsibility represented a sea change in Navy policy and organization for acquisition of surface combatants" (Threston 2009b, p. 109). As a member of the submarine force, I had no personal involvement with this program. However, my father was a senior manager at the Radio Corporation of America (RCA) which served as the prime contractor for Aegis. My father introduced me to Rear Admiral Meyer when I was a junior in high school. I had many occasions to hear firsthand stories about how this amazing Naval engineer was changing the way surface warships were being acquired by the Navy. The admiral's *build a little, test a little, learn a lot* approach was adopted throughout the program and has served me well in my own career. The utilization of system budgets was revolutionary. "In addition to obvious budgets (such as weight, space, power, and cooling), other budgets were established for system errors, reliability, maintainability and availability, system reaction time, maintenance man-hours, and a large number of others" (Threston 2009a, p. 96). The purposeful use and positive effects of systems budgets, which address nonfunctional requirements, has stayed with me and have heavily influenced the notions and structure for this book.

In addition, I have had the opportunity to serve with and work for a number of engineers and technicians who have been tasked with operating and maintaining the systems of Admirals Rickover, Smith, and Meyer. They took the time to teach me the intricacies involved in operating and maintaining complex systems. Special thanks to Art Colling, Stan Handley, Mel Sollenberger, John Almon, Butch Meier, Joe Yurso, Steve Krahn, George Yount, John Bowen, Vince Albano, and Jim Dunn for supporting me in the quest to understand the engineering of systems.

To the many students I have been privileged to teach at both the University of Maryland University College, the College of William and Mary, Virginia Wesleyan College, and Old Dominion University: Your quest for knowledge has challenged me to constantly renew and improve my own understanding as part of the learning process.

To my parents, for providing the love, resources, and basic skills required to thrive as an engineer. To my children, for many hours of challenges, joy, and amazement. Finally, to my wife, for her constant support, companionship, and love throughout the process of completing this book and our journey through life together.

Kevin MacG. Adams

References

Hawkins, W. M. (1994). Levering Smith. In S. Ostrach (Ed.), *Memorial Tributes* (Vol. 7, pp. 214–220). Washington, DC: National Academies Press.

Oliver, D. (2014). *Against the Tide: Rickover's Leadership Principles and the Rise of the Nuclear Navy*. Annapolis, MD: Naval Institute Press.

Threston, J. T. (2009a). The aegis weapons system: Part I. *Naval Engineers Journal, 121*(3), 85–108.

Threston, J. T. (2009b). The aegis weapons system: Part II. *Naval Engineers Journal, 121*(3), 109–132.

Contents

Part III Design Concerns

Part I
Systems Design and Non-functional Requirements

Chapter 1
Introduction to the Design of Engineering Systems

Abstract Engineering Systems, the man-made systems that provide important functions in modern societies, are designed by engineers. The design of engineering systems is a formal process which invokes both technical and human elements to provide the blueprint for systems which simultaneously provide for the needs of a system's stakeholders while not harming the environment or living beings. Engineering design is the term used for the formal processes and methodologies used to create and maintain man-made systems in a life cycle that runs from inception through retirement and disposal.

1.1 Introduction to the Design of Engineering Systems

The term *engineering systems* used in the title of this chapter may be puzzling. This term is used purposefully in order to account for an emerging discipline in engineering that addresses "systems that fulfill important functions in society" (de Weck et al. 2011, p. xi). The term describes "both these systems and new ways of analyzing and designing them" (de Weck et al. 2011, p. xi). The new discipline, engineering systems, addresses technology and technical systems by harmonizing them with the organizational, managerial, policy, political, and human factors that surround the problem while allowing the stakeholder's needs to be met while not harming the larger society. Some may recognize these challenges as being closely related to terminology such as socio-technical systems (STS), engineering-in-the-large, or macro-engineering. Engineering systems is an approach that is in agreement with the author's own holistic worldview for systems endeavors, which invokes a systemic view when dealing with systems, messes, and problems (Hester and Adams 2014).

The chapter will begin by reviewing design and how engineers in the various engineering disciplines are responsible for developing the designs for complex man-made systems. It will conclude by addressing systematic design, the breadth

K.M. Adams, *Non-functional Requirements in Systems Analysis and Design*,
Topics in Safety, Risk, Reliability and Quality 28,
DOI 10.1007/978-3-319-18344-2_1

and depth of disciplines associated with design activities, and the use of life cycle models and supporting processes in the systems design process.

The chapter has a specific learning goal and associated competencies. The learning goal of this chapter is to be able to discriminate engineering design from the less formal design-in-the-small invoked by inventors, hobbyists, entrepreneurs, and ordinary human beings who tend to limit their scope to single, focused purposes, goals, and objectives. This chapter's goal is supported by the following objectives:

- Describe the role of engineering vis-à-vis science.
- Describe the disciplines that affect engineering design.
- Identify the elements that make design a systematic endeavor.
- Describe the 5 major stages in the system life cycle.
- Recognize the technical processes in the system design stage.

The ability to achieve these objectives may be fulfilled by reviewing the materials in the chapter topics which follow.

1.2 Engineering Design

Design is defined as:

> The process of defining the architecture, components, interfaces, and other characteristics of a system or component (IEEE & ISO/IEC 2010, p. 100).

Design is process that can be achieved by inventors, hobbyists, entrepreneurs, and ordinary human beings. However, when the process is applied to complex man-made systems it most often falls under the purview of engineers and the engineering disciplines. 1978 Nobel Laureate Herbert A. Simon [1916–2001] expressed his opinion about engineers and design stating.

> Engineering… are concerned not with the necessary but with the contingent—not with how things are but with how they might be—in short, with design (Simon 1996, p. xii).

As such, a review of some definitions of engineering design is warranted. Table 1.1 contains definitions of engineering design that are worthy of consideration in the quest to understand the uniqueness and importance of the engineering design process.

From the definitions in Table 1.1 it is clear that engineering design is a creative, contingent, problem-solving process that includes technological and social components required for product realization. Those responsible for engineering design are engineers. The section that follows will focus on engineers and engineering in design.

Table 1.1 Definitions for engineering design

Definition	Source
"Engineering design is the process of applying the various techniques and scientific principles for the purpose of defining a device, a process or a system in sufficient detail to permit its physical realization"	Baddour et al. (1961, p. 647)
"Design is a complex process that combines creative thinking, experience, intuition, and quantitative analysis"	Ishii et al. (1988, p. 53)
"Engineering schools to teach about artificial things: how to make artifacts that have desired properties and how to design. Schools of engineering, as well as schools of architecture, business, education, law, and medicine, are all centrally concerned with the process of design"	Simon (1996, p. 111)
"As the key technical ingredient in the product realization process, engineering design bears responsibility for determining in detail how products will be made to meet performance and quality objectives for the customer at a cost that permits a competitive price"	NRC (1991, p. 10)
"Engineering design has both technological and social components. The technological component includes knowledge about engineering science, design methods, engineering models, materials, manufacturing, and computers. The social component includes corporate organization and culture, team design methods, the nature of the design task and of the designer, customer attributes, and employee involvement"	NRC (1991, p. 10)
"An ever-evolving problem-solving activity, engineering design encompasses many different and increasingly advanced practices, including methods for converting performance requirements into product features, computer-integrated manufacturing, cross-functional teams, statistical methods, competitive benchmarking of products, computerized design techniques, and new materials and manufacturing processes. These and other methods used by the most competitive companies worldwide do not exist or operate independently, but rather are integrated into a unified process"	NRC (1991, p. 10)
"Engineering design is always a contingent process, subject to unforeseen complications and influences as the design develops. The precise outcomes of the process cannot be deduced from its initial goal. Design is not, as some textbooks would have us believe, a formal, sequential process that can be summarized in a block diagram"	Ferguson (1994, p. 37)

1.3 Engineers and Engineering in Design

Engineering is "the science by which the properties of matter and the sources of power in nature are made useful to humans in structures, machines, and products" (Parker 1994, p. ix). Theodore von Kármán [1881–1963], the great aeronautical engineering pioneer and co-founder of Cal Tech's Jet Propulsion Laboratory, has been attributed with the following statement.

> The scientist seeks to understand what is. The engineer seeks to create what never was (Petroski 2010, p. 20).

Von Kármán was able to simply state what many committees, boards, and regulators have tried to do in defining engineering. The root for the word engineering is derived from the Latin *ingenium*, which means innate or natural quality. Engineering has many definitions. One of the more comprehensive and thoughtful has been assembled by the historians of engineering (Kirby et al. 1990).

> The art of the practical application of scientific and empirical knowledge to the design and production or accomplishment of various sorts of constructive projects, machines, and materials of use or value to man (p. 2).

Engineering has a number of supporting elements.

> It is customary to think of engineering as part of a trilogy, pure science, applied science, and engineering. It needs emphasis that this trilogy is only one of a triad of trilogies into which engineering fits. The first is pure science, applied science, and engineering. The second is economic theory, finance, and engineering. The third is social relations, industrial relations, and engineering. Many engineering problems are as closely allied to social problems as they are to pure science (Cross 1952, p. 55).

Figure 1.1 is a depiction of engineering in the triad of trilogies.

The *triad of trilogies* helps to characterize the real-world elements that provide the context for engineering endeavors. Engineering applies science to design, develop, and implement solutions. However these solutions exist in the real-world and as such are bounded by financial and social constraints, requiring engineers to have extended knowledge in these disciplines. Dixon (1966) provides the structure

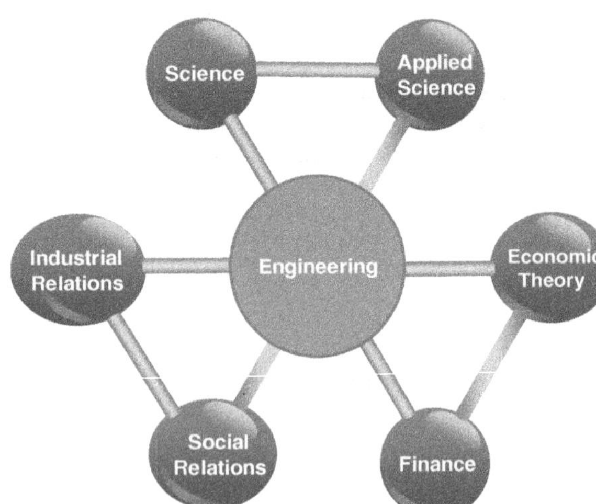

Fig. 1.1 Engineering as an element of the Triad of Trilogies

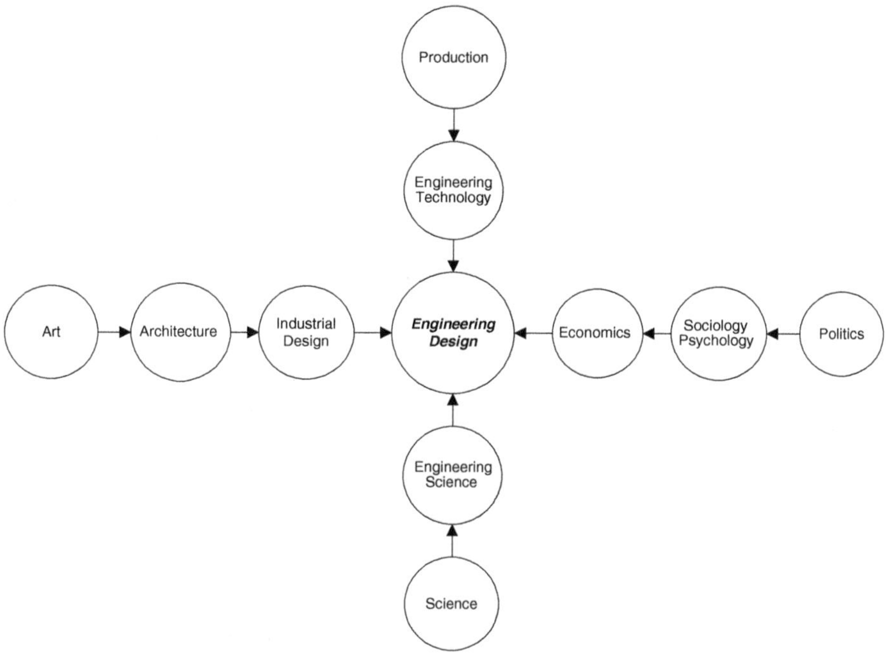

Fig. 1.2 The central activity of engineering design [adapted from a Fig. 1.1 in (Dixon 1966, p. 8)]

for a perspective of the central activity of engineering design in Fig. 1.2 that integrates itself with disciplines such as art, science, politics, and production.

Penny (1970) discusses the principles of engineering design, using Dixon's (1966) perspective and as a starting point for a discussion about the engineering design process.

The process that we label *design* has evolved as human beings have constructed larger systems of increased complexity. A very simple diagram may be used to capture the essential process of engineering design. In Fig. 1.3 engineering design is presented in an ICOM diagram. ICOM is an acronym for Input, Control, Output, and Mechanism (NIST 1993) and is used to describe the functions of the arrow on the processes performed on the system or component in Fig. 1.3.

Each of the elements in Fig. 1.3 are described in terms of one another and the processes used to approach their solution in Table 1.2.

In order to successfully accomplish the tasks in Table 1.2 the engineer is required to acquire and maintain specialized knowledge, skills, and abilities (KSAs). The requisite KSAs are gained through (1) formal education in undergraduate-level engineering programs[1] and (2) during training under the supervision

[1]Engineering programs in the United States are accredited by the Accreditation Board for Engineering and Technology (ABET).

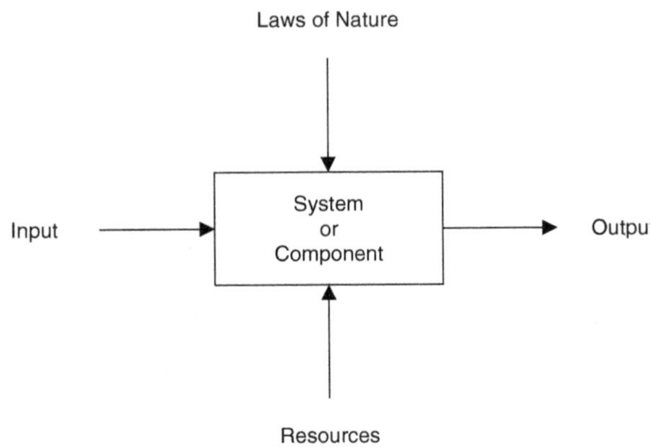

Fig. 1.3 Schematic of engineering design process [adapted from Fig. 1.3 in (Dixon 1966, p. 12)]

Table 1.2 Description of the elements in the engineering design process

Given elements	Solve for	Process
Input, laws of nature, system or component	Output	Analysis (i.e., deduction)
Output, laws of nature, system or component	Input	Inverse analysis (i.e., reverse engineering)
Input, output, system or component	Laws of nature	Science (i.e., induction)
Input, output, laws of nature	System or component	Engineering design

of an experienced engineer.[2] An overview of the KSAs and associated factors that support successful engineering are presented in Table 1.3.

The application of the KSAs required for engineering in a formal process for accomplishing a system design are accomplished through formal design methods. Bayazit (2004) provides a brief historical review of research in design methods in the *systems age* (Ackoff 1974). This area of interest is labeled *design science*.

> Design science uses scientific methods to analyze the structures of technical systems and their relationships with the environment. The aim is to derive rules for the development of these systems from the system elements and their relationships (Pahl et al. 2011, p. 9).

The instantiation of a scientific method for design is a *design methodology* which is defined as:

> A systematic approach to creating a design consisting of the ordered application of a specific collection of tools, techniques, and guidelines (IEEE & ISO/IEC 2010, p. 102).

---

[2]All 50 states have licensing programs where engineers are designated as either *engineers-in-training* or *engineering interns* during the period after they have passed their fundamentals of engineering examination and are serving an apprenticeship. Licensed engineers are designated as Professional Engineers and use the title, P.E.

Table 1.3 KSAs required for successful engineering endeavors

KSA Type	Factor	Definition
Knowledge—"Knowledge refers to an organized body of information associated with facts or procedures which, if applied, makes adequate performance on the job possible" (Chang and Lin 2011, p. 238)	Engineering science	"Thorough knowledge and training in depth in an engineering science specialty" (Dixon 1966, p. 13)
	Manufacturing processes	"Knowledge of and an appreciation for the potential and limitations of both old and new manufacturing processes" (Dixon 1966, p. 13)
Skill—"Skill refers to the proficient manual, verbal or mental manipulation of data or things. Examples of proficient manipulation of things are skill in typing or skill in operating a vehicle" (Chang and Lin 2011, p. 238)	Engineering	Demonstrated proficiency in the application of requisite knowledge and ability during a period of apprenticeship in an engineering science specialty
Ability—"Ability refers to the power to perform an observable activity at the present time. This means that abilities have been evidenced through activities or behaviors that are similar to those required on the job, e.g., ability to plan and organize work" (Chang and Lin 2011, p. 238)	Inventiveness	"The ability to think or discover valuable, useful ideas or concepts" (Dixon 1966, p. 13)
	Engineering analysis	"The ability to analyze a given component, system, or process using engineering or scientific principles in order to arrive quickly at meaningful answers" (Dixon 1966, p. 13)
	Interdisciplinary focus	"The ability to deal competently and confidently with basic problems or ideas form disciplines outside one's specialty" (Dixon 1966, p. 13)
	Computational mathematics	"The ability to bring powerful mathematical and computational skills to a problem when appropriate" (Dixon 1966, p. 13)
	Decision making	"The ability to make decisions in the face of uncertainty but with a fill grasp of all the factors involved" (Dixon 1966, p. 13)
	Communications	"The ability to express oneself clearly and persuasively orally, graphically, and in writing" (Dixon 1966, p. 13)

It is important to note that the word *systematic* is prominent in the definition of a design methodology. Systematic design provides an effective way to *rationalize* the design and production process" (Pahl et al. 2011, p. 9). In developing a rational approach to the design process the design methodology must:

- *allow a problem-directed approach; i.e., it must be applicable to every type of design activity; no matter which specialist field it involves*
- *foster inventiveness and understanding; i.e., facilitate the search for optimum solutions*
- *be compatible with the concepts, methods and findings of other disciplines*
- *not rely on finding solutions by chance*
- *facilitate the application of known solutions to related tasks*
- *be compatible with electronic data processing*
- *be easily taught and learned*
- *reflect the findings or cognitive psychology and modern management science; i.e., reduce workload, save time, prevent human error, and help to maintain active interest*
- *ease the planning and management of teamwork in an integrated and inter-disciplinary product development process*
- *provide guidance for leaders of product development teams* (Pahl et al. 2011, p. 10).

Chapter 2 will discuss a number of methodologies used in engineering design.

1.4 Design in the System Life Cycle Model

The processes that develop, operate, and eventually retire modern systems are best described as a life cycle. The life cycle is a model of the real-world events and processes that surround the system. Life cycle models have distinct phases or stages that mark important points in the life of the system. For systems endeavors life cycle models are described using standards developed by stakeholders from industry, government, and academia that are published by the Institute of Electrical and Electronics Engineers (IEEE) and the International Organization for Standardization (ISO) and the International Electrotechnical Commission (IEC). The standard we use for man-made systems life cycles is *IEEE and ISO/IEC Standard 15288: Systems and software engineering—System life cycle processes* (IEEE & ISO/IEC 2008). Important concepts from IEEE Standard 15288 include:

- Every system has a life cycle. A life cycle can be described using an abstract functional model that represents the conceptualization of a need for the system, its realization, utilization, evolution and disposal.
- The stages represent the major life cycle periods associated with a system and they relate to the state of the system description or the system itself. The stages describe the major progress and achievement milestones of the system through its life cycle. They give rise to the primary decision gates of the life cycle (IEEE & ISO/IEC 2008, p. 10).

A typical systems life cycle model would consist of the stages and associated goals shown in Table 1.4.

Table 1.4 Typical Life Cycle Model stages and goals

Life cycle stage	Goals
Concept	To understand the needs of the system's stakeholders, explore concepts, and develop viable solutions
Design	To transform stakeholder needs to system requirements, to create solution descriptions, build the system, and to verify and validate the system
Production	To produce, inspect and test the system
Operations and Maintenance	To operate and maintain the system
Retirement and Disposal	To replace and responsibly dispose of the existing system

The design stage of the system life cycle uses a number of technical processes described in IEEE Standard 15288 (2008) to accomplish the goals of design. Table 1.5 is a listing of the technical processes and the purposes that are invoked to accomplish the design phase.

The detailed outcomes, and associated activities and tasks for each technical process are described in IEEE Std 15288 (2008).

Table 1.5 Technical Processes in the Design Stage

Technical Process	Purpose
Stakeholder requirements definition	"define the requirements for a system that can provide the services needed by users and other stakeholders in a defined environment" (IEEE & ISO/IEC 2008, p. 36)
Requirements Analysis	"transform the stakeholder, requirement-driven view of desired services into a technical view of a required product that could deliver those services" (IEEE & ISO/IEC 2008, p. 39)
Architectural Design	"synthesize a solution that satisfies system requirements" (IEEE & ISO/IEC 2008, p. 40)
Implementation	"realize a specified system element" (IEEE & ISO/IEC 2008, p. 43)
Integration	"assemble a system that is consistent with the architectural design" (IEEE & ISO/IEC 2008, p. 44)
Verification	"confirm that the specified design requirements are fulfilled by the system" (IEEE & ISO/IEC 2008, p. 45)
Transition	"establish a capability to provide services specified by stakeholder requirements in the operational environment" (IEEE & ISO/IEC 2008, p. 46)
Validation	"provide objective evidence that the services provided by a system when in use comply with stakeholders' requirements, achieving its intended use in its intended operational environment" (IEEE & ISO/IEC 2008, p. 47)

1.5 Summary

The information presented in this chapter introduced engineers and how they transform human needs, through formal methods of design, into complex man-made systems. The systematic nature of design and the breadth and depth of disciplines associated with design activities were also reviewed. The systematic nature of engineering and the use of life cycle models and supporting processes were highlighted as essential elements of engineering design.

The next chapter will discuss a number of engineering methodologies that may be used to invoke repeatable processes for the engineering design of man-made systems.

References

Ackoff, R. L. (1974). The systems revolution. *Long Range Planning, 7*(6), 2–20.

Baddour, R. F., Holley, M. J., Koppen, O. C., Mann, R. W., Powell, S. C., Reintjes, J. F., et al. (1961). Report on engineering design. *Journal of Engineering Education, 51*(8), 645–661.

Bayazit, N. (2004). Investigating design: A review of forty years of design research. *Design Issues, 20*(1), 16–29.

Chang, H.-L., & Lin, J.-C. (2011). Factors that Impact the Performance of e-Health Service Delivery System. In: *Proceedings of the 2011 International Joint Conference on Service Sciences (IJCSS)* (pp. 237–241). Los Alamitos, CA: IEEE Computer Society.

Cross, H. (1952). *Engineers and Ivory Towers*. New York: McGraw-Hill.

de Weck, O. L., Roos, D., & Magee, C. L. (2011). *Engineering systems: Meeting human needs in a complex technological world*. Cambridge, MA: MIT Press.

Dixon, J. R. (1966). *Design engineering: Inventiveness, analysis, and decision making*. New York: McGraw Hill.

Ferguson, E. S. (1994). *Engineering and the Mind's Eye*. Cambridge, MA: MIT Press.

Hester, P. T., & Adams, K. M. (2014). *Systemic thinking—Fundamentals for understanding problems and messes*. New York: Springer.

IEEE, & ISO/IEC. (2008). *IEEE and ISO/IEC Standard 15288: Systems and software engineering —System life cycle processes*. New York and Geneva: Institute of Electrical and Electronics Engineers and the International Organization for Standardization and the International Electrotechnical Commission.

IEEE, & ISO/IEC. (2010). *IEEE and ISO/IEC Standard 24765: Systems and Software Engineering —Vocabulary*. New York and Geneva: Institute of Electrical and Electronics Engineers and the International Organization for Standardization and the International Electrotechnical Commission.

Ishii, K., Adler, R., & Barkan, P. (1988). Application of design compatibility analysis to simultaneous engineering. *Artificial Intelligence for Engineering Design, Analysis and Manufacturing, 2*(1), 53–65.

Kirby, R. S., Withington, S., Darling, A. B., & Kilgour, F. G. (1990). *Engineering in history*. Mineola, NY: Dover Publications.

NIST. (1993). *Integration Definition for Function Modeling (IDEF0) (FIPS Publication 183)*. Gaithersburg, MD: National Institute of Standards and Technology.

NRC. (1991). *Improving engineering design: Designing for competitive advantage*. Washington, DC: National Academy Press.

Pahl, G., Beitz, W., Feldhusen, J., & Grote, K.-H. (2011). *Engineering design: A systematic approach (K. Wallace & L. T. M. Blessing, trans)* (3rd ed.). Darmstadt: Springer.

Parker, S. (Ed.). (1994). *McGraw-Hill Dictionary of eEngineering*. New York: McGraw-Hill.

Penny, R. K. (1970). Principles of engineering design. *Postgraduate Medical Journal, 46*(536), 344–349.

Petroski, H. (2010). *The essential engineer: Why science alone will not solve our global problems*. New York: Alfred A. Knopf.

Simon, H. A. (1996). *The sciences of the artificial* (3rd ed.). Cambridge, MA: MIT Press.

Chapter 2
Design Methodologies

Abstract Engineering design is a formal discipline within the field of engineering. The study of design methodologies is a sub-discipline and requires the use of unique modes of thought and the application of a number of specific features to ensure that designs are both repeatable and result in products that are useful for a specified period of service. A methodology is purposefully positioned in a formal hierarchy of scientific approaches, supported by a specific paradigm and philosophy while acting as the framework for more detailed methods and techniques. There are a number of unique engineering design methodologies, frameworks, and models that have evolved to provide the structural framework for the applicable design processes, methods, and techniques. The Axiomatic Design Methodology provides a systems-based framework for design that permits design alternatives to be evaluated based on quantitative analysis, eliminating the need for messy qualitative and cost-based models.

2.1 Introduction to Design Methodologies

This chapter will introduce a number of engineering methodologies that may be used to invoke repeatable processes for the purposeful design of *engineering systems*. The term *engineering systems* may be used either as (1) a noun—"systems that fulfill important functions in society" (de Weck et al. 2011, p. xi) or (2) a verb—"new ways of analyzing and designing them" (de Weck et al. 2011, p. xi). In the verb form *engineering systems* addresses technology and technical systems by harmonizing them with the organizational, managerial, policy, political, and human factors that surround the problem while allowing the stakeholder's needs to be met while not harming the larger society. The analysis and design efforts for engineering systems require formal methodologies in order to implement repeatable processes that both invoke proven engineering processes and are subject to efforts to improve those processes.

The chapter will begin by discussing the discipline of engineering design and its sub-disciplines of design theory and design methodology. The features and modes of thought that support engineering design endeavors are reviewed.

K.M. Adams, *Non-functional Requirements in Systems Analysis and Design*,
Topics in Safety, Risk, Reliability and Quality 28,
DOI 10.1007/978-3-319-18344-2_2

The next section defines the terminology required to understand how a methodology is positioned in the scientific paradigm. The section that follows provides a formal hierarchical relationship between the terms.

This is followed by a section that presents seven historically significant engineering design methodologies. The basics tenets of each methodology, including the major phases, stages, and steps associated with each model are reviewed. References for further study of each methodology are provided.

The chapter concludes by presenting a formal methodology for accomplishing engineering technical processes, the Axiomatic Design Methodology. The Axiomatic Design Methodology provides the framework through which system functional and non-functional requirements are satisfied by design parameters and process variables in the system design.

The chapter has a specific learning goal and associated competencies. The learning goal of this chapter is to be able to identify describe engineering design, its position in the scientific paradigm and a number of specific methodologies for conducting engineering design endeavors. This chapter's goal is supported by the following objectives:

- Describe engineering design as a discipline.
- Differentiate between design theory and design methodology.
- Describe the desirable features of engineering design.
- Describe the double-diamond model of design.
- Construct a hierarchical structure that includes a paradigm, philosophy, methodology method, and technique.
- Differentiate between the seven historical design methodologies.
- Describe the major features of the Axiomatic Design Methodology.

The ability to achieve these objectives may be fulfilled by reviewing the materials in the chapter topics which follow.

2.2 Introduction to the Discipline of Engineering Design

The study of engineering design is a discipline within the broader field of engineering. The scholarly journals in the discipline that address transdisciplinary engineering design topics are presented in Table 2.1.

Design theory (or design science) and design methodology represent two academic subjects within the discipline of engineering design that each have their own unique viewpoints and research agendas. The two subject areas are simply defined as follows:

- Design theory—*is descriptive as indicates what design is* (Evbuomwan et al. 1996, p. 302).
- Design methodology—*is prescriptive as it indicates how to do design* (Evbuomwan et al. 1996, p. 302)

Table 2.1 Scholarly journals for engineering design

Journal (ISSN)	Description	Publication period/issues
Journal of Engineering Design (0954–4828)	Articles on research into the improvement of the design processes/practices in industry and the creation of advanced engineering products, and academic studies on design principles	1990-present, 4/year
Research in Engineering Design (0934–9839)	Research papers on design theory and methodology in all fields of engineering, focusing on mechanical, civil, architectural, and manufacturing engineering	1989-present, 4/year
Design Issues (0747–9360)	Examines design history, theory, and criticism. Provokes inquiry into the cultural and intellectual issues surrounding design	1984-present, 4/year
Design Studies (0142–694X)	Approaches the understanding of design processes from comparisons across all domains of application, including engineering and product design, architectural and urban design, computer artefacts and systems design	1979-present, 6/year
Journal of Mechanical Design (1050–0472)	Serves the broad design community as the venue for scholarly, archival research in all aspects of the design activity with emphasis on design synthesis	1978-present, 12/year
Journal of Research Design (1748–3050)	Interdisciplinary journal, emphasizing human aspects as a central issue of design through integrative studies of social sciences and design disciplines	2001-present, 4/year

This chapter is focused upon design methodology—the *how to* of design. More precisely, how does a specific engineering design methodology sequence and execute the technical processes of the design stage of the systems life cycle which were described at the end of Chap. 1? The purpose of each of the technical processes required for the design stage are established in *IEEE Standard 15288* (2008) and presented in Table 2.2 (repeated from Table 1.5).

The section that follows will discuss the features and properties that a design methodology must possess in order to effectively execute the eight processes in Table 2.2.

2.2.1 *Features that Support Design Methodologies*

> Self conscious design contains many well-known activities such as decision making, optimization, modeling, knowledge production, prototyping, ideation, or evaluation. However, it cannot be reduced to any one of them or all of these activities (e.g., decisions are made in design, but design is more than decision making). Thus, design theory is not about modeling everything that one can find in a design practice, its goal is to precisely

Table 2.2 Technical processes in the design stage

Technical process	Purpose
1. Stakeholder requirements definition	"Define the requirements for a system that can provide the services needed by users and other stakeholders in a defined environment" IEEE and ISO/IEC (2008, p. 36)
2. Requirements analysis	"Transform the stakeholder, requirement-driven view of desired services into a technical view of a required product that could deliver those services" IEEE and ISO/IEC (2008, p. 39)
3. Architectural design	"Synthesize a solution that satisfies system requirements" IEEE and ISO/IEC (2008, p. 40)
4. Implementation	"Realize a specified system element" IEEE and ISO/IEC (2008, p. 43)
5. Integration	"Assemble a system that is consistent with the architectural design" IEEE and ISO/IEC (2008, p. 44)
6. Verification	"Confirm that the specified design requirements are fulfilled by the system" IEEE and ISO/IEC (2008, p. 45)
7. Transition	"Establish a capability to provide services specified by stakeholder requirements in the operational environment" IEEE and ISO/IEC (2008, p. 46)
8. Validation	"Provide objective evidence that the services provided by a system when in use comply with stakeholders' requirements, achieving its intended use in its intended operational environment" IEEE and ISO/IEC (2008, p. 47)

> address issues that are beyond the scope of the classical models that accompany its constituent activities (decision making, prescriptive models, hypothetic-deductive model and others). The questions this goal raises are of course: What, then, are the core phenomena of Design? Is Design driven by novelty, continuous improvement, creativity, or imagination? (Le Masson et al. 2013, pp. 97–98).

There are a number of features (or properties) that should be possessed by each and every design methodology. The features are prominent elements characteristic of each and every successful engineering design endeavor. While most design methodologies do not formally address these features, they are unwritten elements that the both the methodology and team members must invoke during the design endeavor. The features are foundational to every engineering design methodology and ensure that the methodology effectively and efficiently executes the eight technical processes of the design stage. Ten features that support design methodologies are presented in Table 2.3 (Evbuomwan et al. 1996).

All of the features in Table 2.3 represent unique facets (i.e., one side of something many-sided) that the design methodology must contain in order to effectively and efficiently execute the technical processes of the design stage during the systems life cycle. The first letters of each of the features may be combined to create the acronym *ERICOIDITI*. The features are depicted in Fig. 2.1.

The next section will address the types of thought invoked during the execution of a design methodology.

Table 2.3 Desirable features that support design methodologies

Feature	Description
Exploratory	Design is a formal professional endeavor requiring specific knowledge, skills, and abilities
Rational	Design is rational involving logical reasoning, mathematical analysis, computer simulation, laboratory experiments and field trials, etc
Investigative	Design requires inquiry into the stakeholder's requirements and expectations, available design techniques, previous design solutions, past design failures and successes, etc
Creative	Design requires know-how, ingenuity, memory, pattern recognition abilities, informed solution scanning, lateral thinking, brainstorming, analogies, etc
Opportunistic	Both top-down and bottom-up approaches are used by the design team based upon the situation presented
Incremental	Improvements or refinements are proposed during the design process in order to achieve an improved design
Decision-making	Design requires value judgements. Courses of action and selection from competing solutions are based on experience and criteria provided by the system's stakeholders
Iterative	Design is iterative. Artifacts are analyzed with respect to functional and non-functional requirements, constraints, and cost. Revisions are based on experience and feedback mechanisms
Transdisciplinary	Design of engineering systems requires a transdisciplinary team
Interactive	Design is interactive. The design team is an integral part of the actual design

2.2.2 *Thought in a Design Methodology*

Design teams invoke different modes of thinking during execution of a design methodology. The particular type of thought is a function of the execution point in the methodology and the unique process being utilized. Two major modes of thought exist in most methodologies—divergence and convergence, which are inter-related and complementary. The sequence of divergence and convergence is represented by the *Double-diamond Model of Design* (Norman 2013) depicted in Fig. 2.2.

The idea behind the Double-diamond is that when a design idea is conceived, the first action is to expand the team's thinking (divergence) allowing the team to explore all of the issues related to the design idea. Once all of the ideas related to the design idea are surfaced and reviewed, the team may then focus their thinking (convergence) on what the design should do. After the team has decided what the design should accomplish, the team must once again expand their thinking (divergence) to review all of the possible solution alternatives for the system. Finally, once all of the solution alternatives have been identified and reviewed, the team may focus (convergence) on a single satisficing solution.

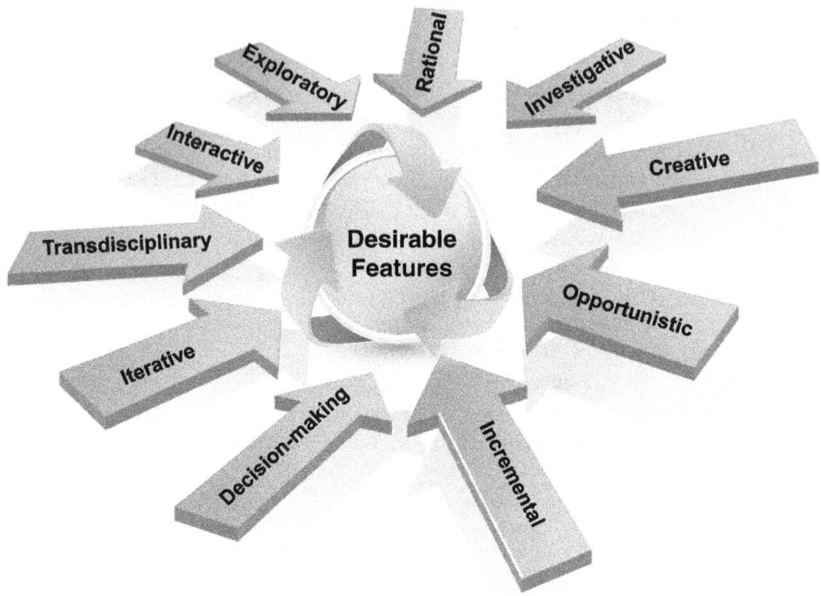

Fig. 2.1 Desirable features of engineering methodologies

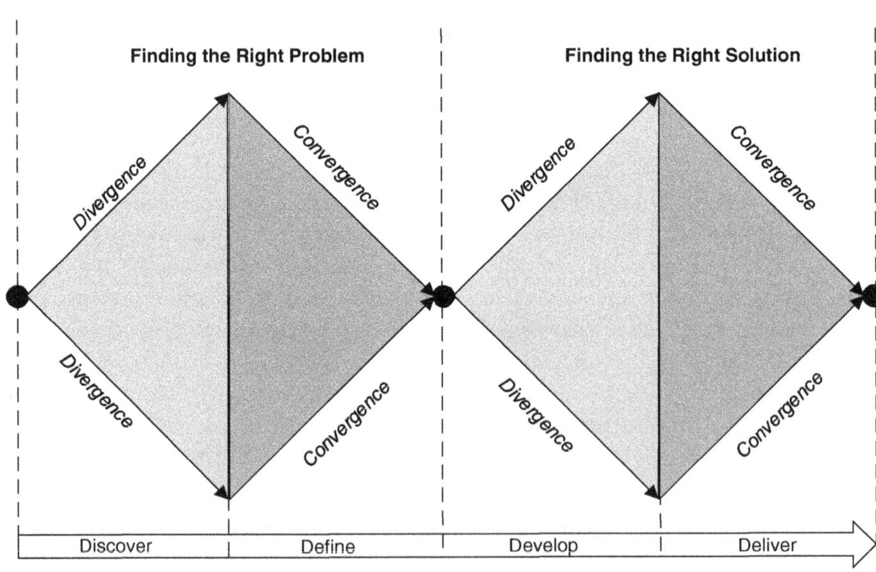

Fig. 2.2 Double diamond model of design

2.2.3 Synthesis of Thought and Features that Support All Engineering Design Methodologies

To be successful, the design team must invoke both the desirable features and the two modes of thinking as a matter of routine during the execution of the technical processes within the design methodology adopted for the design endeavor. The ability to apply thinking modes required for the technical processes and to include each of the desirable features provides a solid framework for the design effort. Figure 2.3 is a depiction of the synthesis of thought and desirable features that provide support for all engineering design methodologies.

The section that follows will review the terminology and relationships associated with a methodology.

2.3 Methodological Terms and Relationships

In order to better understand where a methodology fits in the scientific hierarchy, the next section will review and define key terms that include paradigm, philosophy, methodology, method, and technique. The section will also present a structured relationship between the terms.

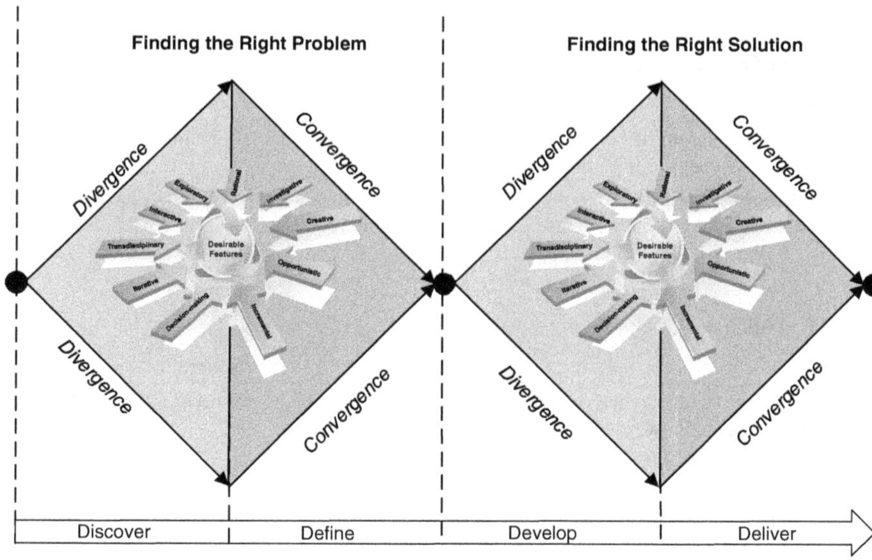

Fig. 2.3 Synthesis of features and thought

2.3.1 Paradigm

Paradigm is a term attributed to the physicist and modern Philosopher of Science Thomas Kuhn [1922–1996]. Two definitions for paradigm are presented in Table 2.4.

From these definitions a composite definition for paradigm is developed which states that a paradigm is *the whole network of theories, beliefs, values, methods, objectives, professional and educational structure of a scientific community*. As such the paradigm associated with the field of engineering and the discipline of engineering design requires the paradigm to contain:

- The network of beliefs and values,
- The professional and education structure, and
- The worldview for a scientific community.

2.3.2 Philosophy

Two relevant definitions of philosophy are presented in Table 2.5.

From these definitions a composite definition for philosophy states that philosophy is *the logical analysis of the concepts, propositions, proofs, theories of science, as well as of those which we select in available science as common to the*

Table 2.4 Definitions of paradigm

Definition	Source
"The entire constellation of beliefs, values, techniques, and so on shared by the members of a given community"	Kuhn (1996, p. 175)
"The whole network of theories, beliefs, values, methods, objectives, professional and educational structure of a scientific community"	Psillos (2007, p. 174)

Table 2.5 Definitions of philosophy

Definition	Source
"Philosophy deals with science only from the logical viewpoint. Philosophy is the logic of science, i.e., the logical analysis of the concepts, propositions, proofs, theories of science, as well as of those which we select in available science as common to the possible methods of constructing concepts, proofs, hypotheses, theories"	Carnap (1934, p. 6)
"The different ways in which we classify whatever the world, or any world, contains"	Proudfoot and Lacey (2010, p. 302)

possible methods of constructing concepts, proofs, hypotheses, theories. The application of the notion of *philosophy* to the field of engineering and discipline of engineering design requires the supporting philosophy to contain:

- The body of theoretical knowledge that underpins the worldview,
- Is at the highest level of abstraction, and
- Contains the systems laws, principles, theorems, and axioms used by the scientific community to address the world.

2.3.3 Methodology

Three relevant definitions of a methodology are presented in Table 2.6.

From these definitions a composite definition for methodology states that a methodology is *the systematic analysis and organization of the rational and experimental principles and processes which guide a scientific inquiry*. The application of the definition for methodology to the field of engineering and the discipline of engineering design requires the supporting methodology to be:

- It is one of the systemic approaches that is used to guide scientific endeavor, and
- Is a blend of more than one systemic methodology, becoming increasingly specific until it becomes a unique methodology.

2.3.4 Method or Technique

Both method and technique are terms that require definition in order to both differentiate them and provide for a common language for engineering design.

- **Method**: A systematic procedure, technique, or mode of inquiry employed by or proper to a particular discipline or art (Mish 2009, p. 781).
- **Technique**: A body of technical methods (as in a craft or in scientific research) (Mish 2009, p. 1283).

Table 2.6 Definitions of methodology

Definition	Source
"The philosophical study of the scientific method."	Honderich (2005, p. 598)
"A structured set of methods or techniques to assist people in undertaking research or intervention."	Mingers (2003, p. 559)
"The systematic analysis and organization of the rational and experimental principles and processes which guide a scientific inquiry, or which constitute the structure of the special sciences more particularly."	Runes (1983, p. 212)

The section that follows will provide a hierarchical relationship between each of the terms utilized in the description of an engineering methodology.

2.3.5 Relationship Between Scientific Terms

There is a distinct relationship between the terms paradigm, philosophy, methodology, method, and technique. Dr. Peter Checkland, acknowledged as a leader in the development of systemic methodologies states:

> I take a methodology to be intermediate in status between a philosophy, and a technique … a technique is a precise specific programme of action which will produce a standard result … A methodology will lack the precision of a technique but will be a firmer guide to action than a philosophy (Checkland 1999, p. 162).

The relationships between a paradigm, philosophy, methodology, method, and technique are depicted in Fig. 2.4.

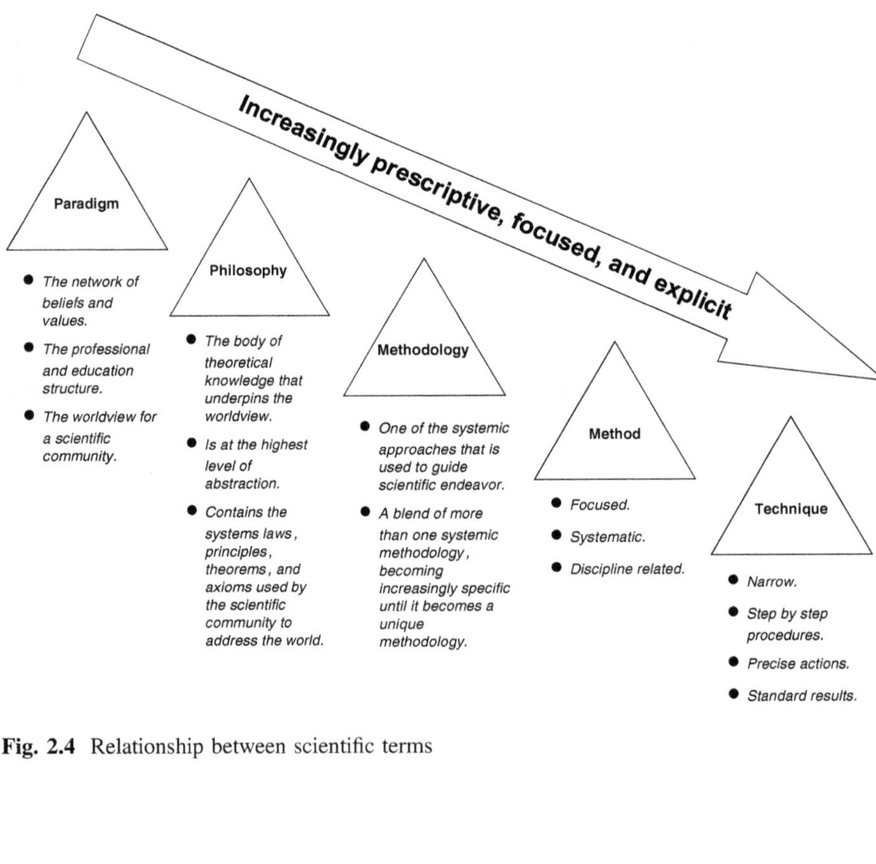

Fig. 2.4 Relationship between scientific terms

2.4 Hierarchical Structure for Engineering Design

Figure 2.4 provides the structure within which an engineering design methodology exists. Documents that support a methodology in engineering design will be discussed in the section that follow.

2.4.1 *Paradigm for Engineering as a Field of Science*

The top-level, the paradigm, that surrounds all engineering efforts is science, the scientific method and scientific community. The "sciences are organized bodies of knowledge" (Nagel 1961, p. 3) which at its highest level includes six major fields: (1) natural sciences; (2) engineering and technology; (3) medical and health sciences; (4) agricultural sciences; (5) social sciences; and (6) humanities (OECD 2007). Each science is guided by "the desire for explanation which are at once systematic and controllable by factual evidence that generates science" (Nagel 1961, p. 4). "Science is not a rigid body of facts. It is a dynamic process of discovery. It is alive as live itself" (Angier 2007, p. 19).

2.4.2 *Philosophy for Engineering*

The second-level, philosophy, serves to focus all engineering efforts and contains a guide to the theoretical body of knowledge that underpins the worldview for all engineers. There is an overarching body of knowledge that encompasses general engineering knowledge (NSPE 2013) and individual bodies of knowledge for each engineering discipline. For instance, the *Guide to the Systems Engineering Body of Knowledge* or SEBoK (BKCASE-Editorial-Board 2014) and the *Guide to the Software Engineering Body of Knowledge* or SWEBOK (Bourque and Fairley 2014). The body of knowledge acts as a guide to the specific knowledge areas required to effectively practice engineering in the discipline governed by the body of knowledge. Each body of knowledge endeavors to:

- To promote a consistent worldwide view of the engineering discipline,
- To specify the scope of, and clarify the relationship of the engineering discipline with other scientific fields and engineering disciplines,
- To characterize the contents of the engineering discipline,
- To provide a topical access to the body of knowledge in the extant literature, and
- To provide a foundation for curriculum development and for individual certification and licensing material in the discipline.

2.4.3 *Methodology for Engineering Design*

The third level, the methodology, serves to focus all engineering design efforts (a discipline of engineering) in achieving the technical processes required to design a man-made systems. The definition of a design methodology was provided in Chap. 1.

> A systematic approach to creating a design consisting of the ordered application of a specific collection of tools, techniques, and guidelines (IEEE and ISO/IEC 2010, p. 102).

A design methodology can be envisioned as a framework or model that focuses the actions of human beings that are attempting to define an object, device, process, or system in order to provide the details required to effect construction, assembly, and implementation for use.

> Design models are the representations of philosophies or strategies proposed to show how design is and may be done (Evbuomwan et al. 1996, p. 305).

The section that follows will present a number of formal methodologies that may be utilized during the design of engineering system.

2.5 Engineering Design Methodologies

This section will present a number of major engineering design methodologies. Each methodology will be reviewed at a very high-level, but will contain adequate references that may guide further investigation of the details of the methodology. It is important to remember that the methodology is a framework or model that guides the execution, tracking, and accomplishment of technical tasks required to accomplish the design of a man-made system. Seven select methodologies will be presented in chronological order of their appearance in the literature.

2.5.1 *Methodology by Morris Asimov*

Morris Asimow [1906–1982], a professor of engineering systems at the University of California at Los Angeles, developed the seven-phase linear chronological structure (i.e., morphology) for design projects depicted in Fig. 2.5. Asimow was the initial author to discuss morphology in engineering design and as such has the distinction of authoring one of the earliest texts on the topic.

Notice that Fig. 2.5 contains three phases identified as the design phases. The purposes of each design phase are as follows:

- Feasibility study—"to achieve a set of useful solutions to the design problem" (Asimow 1962, p. 12).

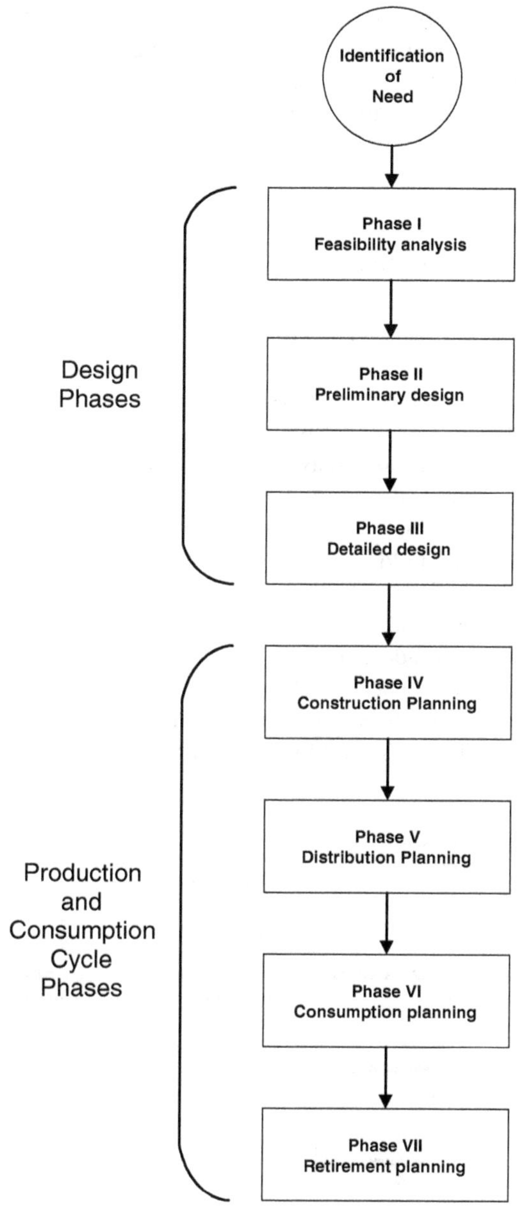

Fig. 2.5 Seven phases of a complete project [adapted from (Asimow 1962, p. 12)]

- Preliminary design—"to establish which of the proffered alternatives is the best design concept" (Asimow 1962, p. 13).
- Detailed design—"to furnish the engineering description of a tested and producible design" (Asimow 1962, p. 13)

This step-wise execution of design phases and the associated processes should be familiar to every engineer as it serves as the foundation for teaching the sequential path of activities involved in delivering products and systems. The details of each of the seven phases are presented as individual chapters in Asimow's text *Introduction to Design* (1962).

2.5.2 Methodology by Nigel Cross

Nigel Cross, an emeritus professor of design studies at The Open University in the United Kingdom and current editor-in-chief of the scholarly journal *Design Studies*, published the first version of the eight-stage model of design shown in Fig. 2.6 in 1984. This model is unique in that it permits the user to visualize how the larger design problem may be broken into sub-problems and sub-solutions which are then synthesized into the total solution.

The three stages on the left hand side of the model and the one stage in the bottom middle establishes objectives, functions, requirements, and characteristics of the problem. The three stages on the right hand side of the model and the one in the upper middle generate, evaluate, and provide improvements to alternatives and identify additional opportunities that may be relevant to the problem's design solution. The right hand side responds to and provides feedback to the left hand side. The details of this model are presented in Cross' text *Engineering Design Methods* (Cross 2008), which is now in its 4th edition.

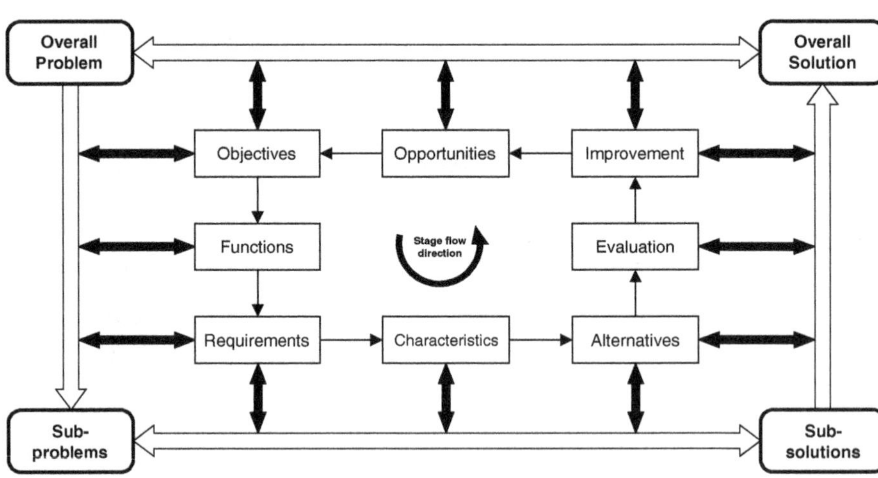

Fig. 2.6 Eight stages of the design process [adapted from (Cross 2008, p. 57)]

2.5.3 Methodology by Michael J. French

Michael J. French, Emeritus Professor of Engineering Design at Lancaster University proposed a block diagram of total design in his text *Conceptual Design for Engineers* in 1985. His four-phase model is depicted in Fig. 2.7.

The details associated with this model of design are presented in his text *Conceptual Design for Engineers* (French 1998) which has recently been reissued in its 3rd edition.

Fig. 2.7 Block diagram of the design process [adapted from Fig. 1.1 in (French 1998, p. 2)]

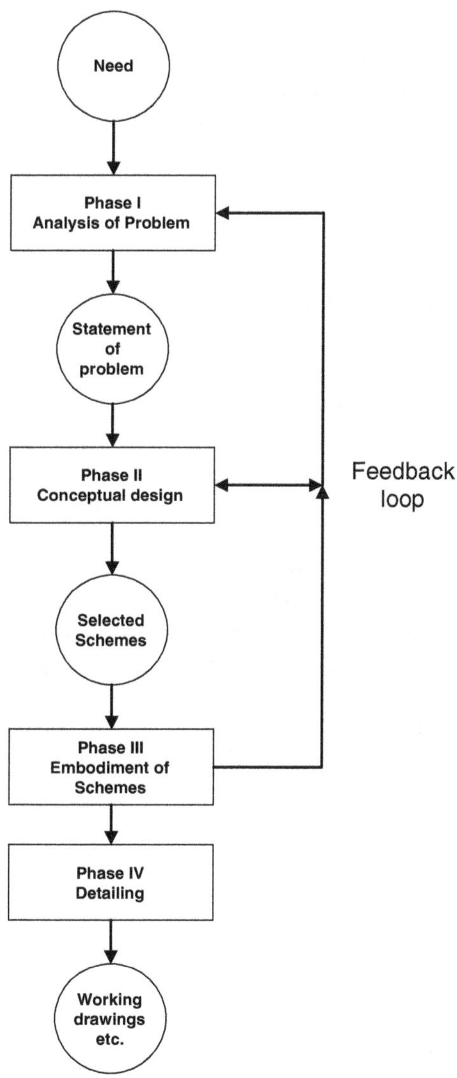

2.5.4 Methodology by Vladimir Hubka and W. Ernst Eder

Vladimir Hubka [1924–2006] was the head of design education at the Swiss Federal Technical University (ETH) in Zürich from 1970 until 1990. His area of expertise was design science and the theory of technical systems. Hubka proposed a four-phase, six step model that addressed elements of design from concept through creation of assembly drawings. A simplified depiction of Hubka's design process model that represents the states of the technical processes during the design phases is depicted in Fig. 2.8.

This is a unique model in that specific design documents are identified as deliverable objects upon completion of the various steps. The details of this innovative approach to design are described in the text *Design Science: Introduction to the Needs, Scope and Organization of Engineering Design Knowledge* (Hubka and Eder 1995). Hubka's long-time colleague W. Ernst Eder provides an excellent compilation on Hubka's legacy, which includes his views on both engineering design science and the theory of technical systems, providing a glimpse into a number of fascinating views on these subjects (Eder 2011).

2.5.5 Methodology by Stuart Pugh

Stuart Pugh [1929–1993] was the Babcock Professor of Engineering Design and the head of the Design Division at the University of Strathclyde in Glasgow, Scotland from 1985 until him untimely death in 1993. During his time at Strathclyde he completed his seminal work *Total Design: Integrated Methods for Successful Product Engineering* (1991). Pugh was an advocate of participatory design using transdisciplinary teams. Until Pugh fostered this idea in both his teaching and consulting work, most engineers focused on technical elements of the design and rarely participated in either the development process or the commercial aspects associated with the product. Pugh's use of transdisciplinary teams ensured that both technical and non-technological factors were included in what he labeled *Total Design*.

Pugh's *Total Design Activity Model* has four parts. The first part is the *design core* of six phases: (1) user need; (2) product specification; (3) conceptual design; (4) detail design; (5) manufacture; and (6) and sales. The six phases of the design core are depicted in Fig. 2.9. The iterations between the phases account for changes to the objectives for the product during the period of design.

The second part of the *Total Design Activity Model* is the product design specification (PDS). The PDS envelopes the design core and contains the major specification elements required to design, manufacture and sell the product. The major elements of a PDS are presented in Table 2.7.

When the PDS is placed on the design core the *Total Design Activity Model* is represented by two of its four parts as depicted in Fig. 2.10. The lines that radiate

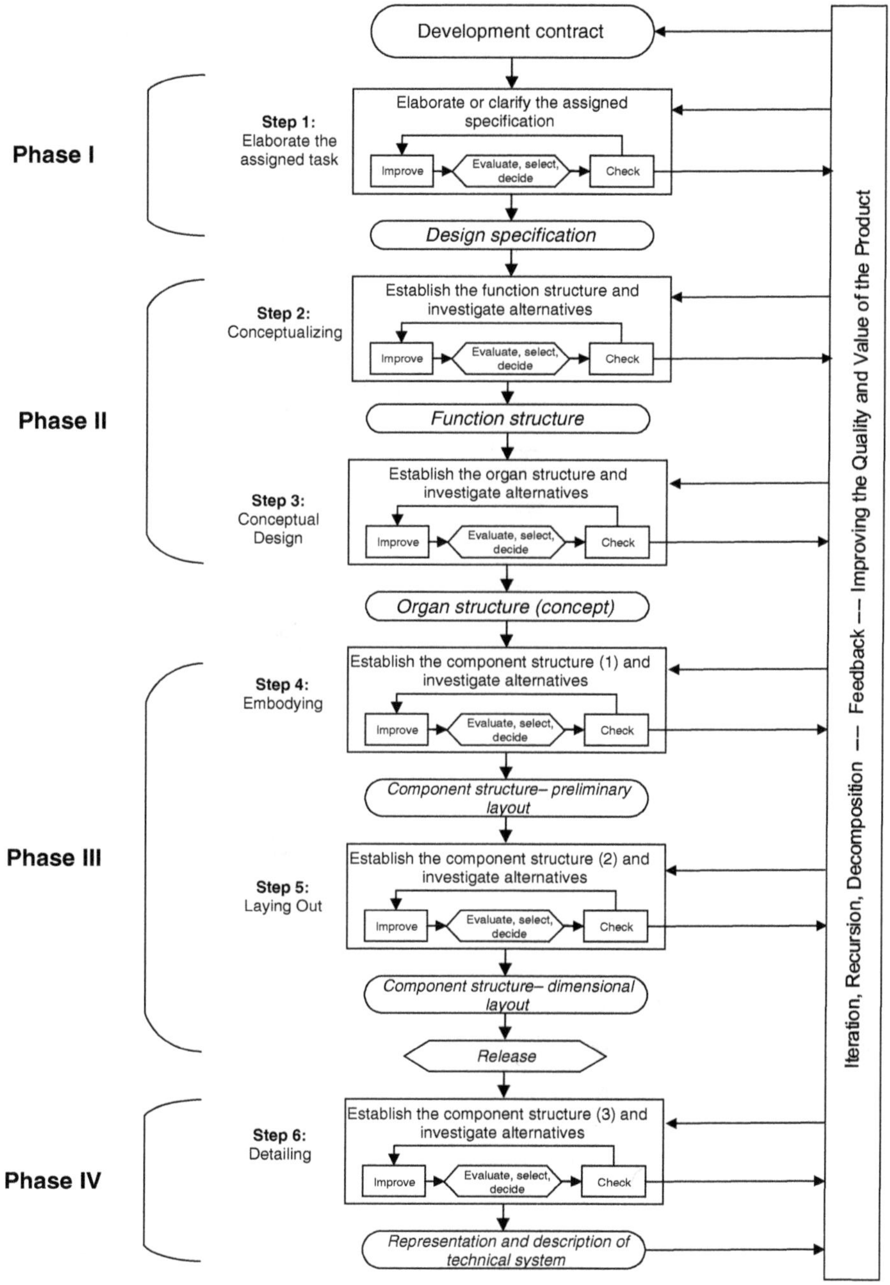

Fig. 2.8 Depiction of Hubka's design model [adapted from Figs. 7–13 (Hubka and Eder 1995)]

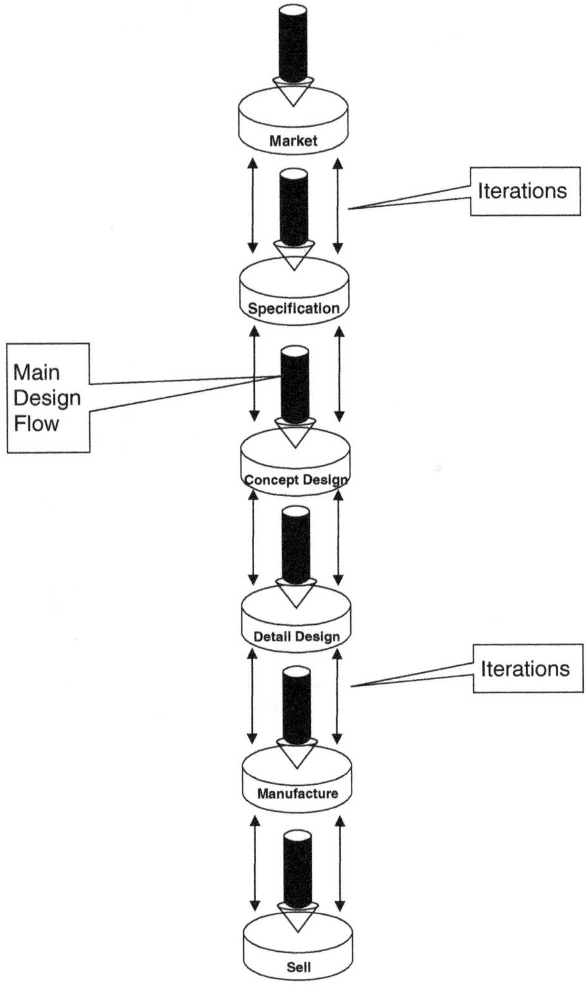

Fig. 2.9 Main design core [adapted from Fig. 1.4 in (Pugh 1991, p. 6)]

Table 2.7 Elements of the product design specification

Customer	Processes	Size	Shipping	Performance
Disposal	Aesthetics	Politics	Installation	Weight
Maintenance	Competition	Packing	Reliability	Shelf life storage
Patents	Environment	Testing	Safety	Legal
Documentation	Quality	Product lifespan	Materials	Ergonomics
Standards specifications	Manufacturing facilities	Market constraints	Company constraints	Life in service
Product cost	Time scale			

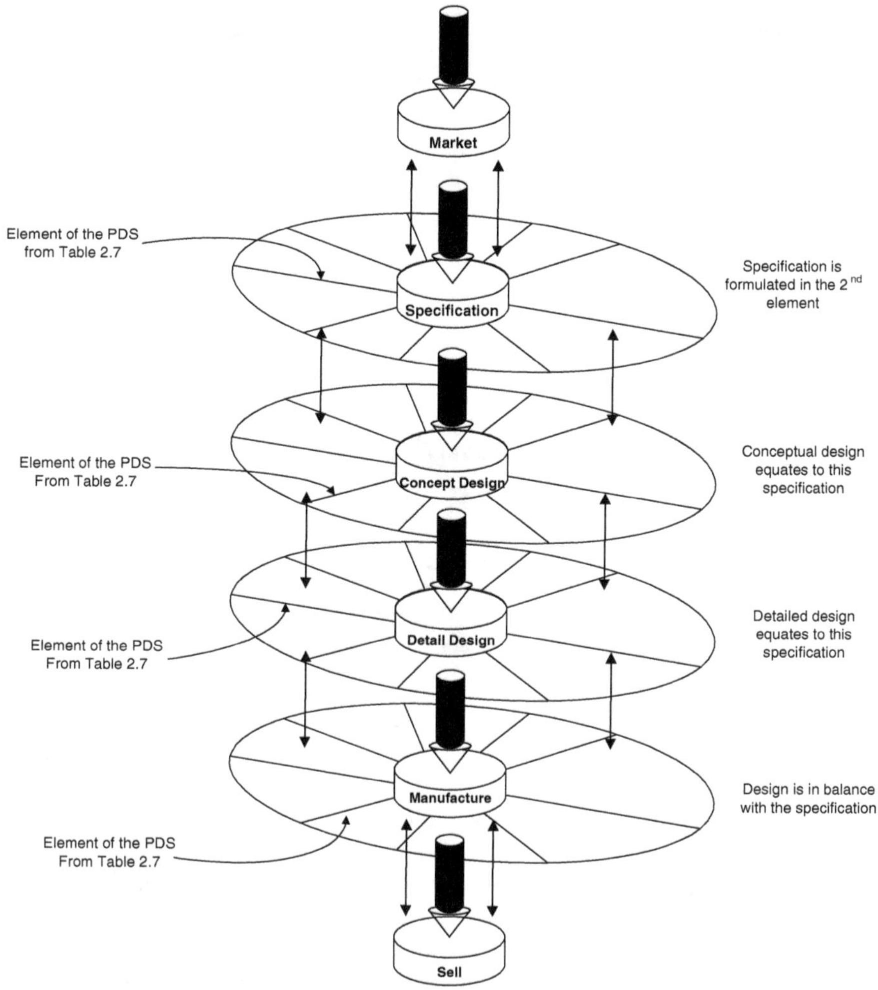

Fig. 2.10 Design core and surrounded by PDS [adapted from Fig. 1.5 in (Pugh 1991, p. 7)]

from and surround the core phases are the elements of the PDS relevant to the particular product's design.

The third part of the *Total Design Activity Model* are the inputs from the discipline independent methods required to execute the design core. These include both the desirable features of engineering design and the two modes of thought as depicted in Fig. 2.3 and many others. The fourth and final part of the *Total Design Activity Model* are the inputs from the technology and discipline dependent sources. Many discipline specific methods are required to execute the elements of the PDS that surround the design core. Examples include stress and strain analysis, welding,

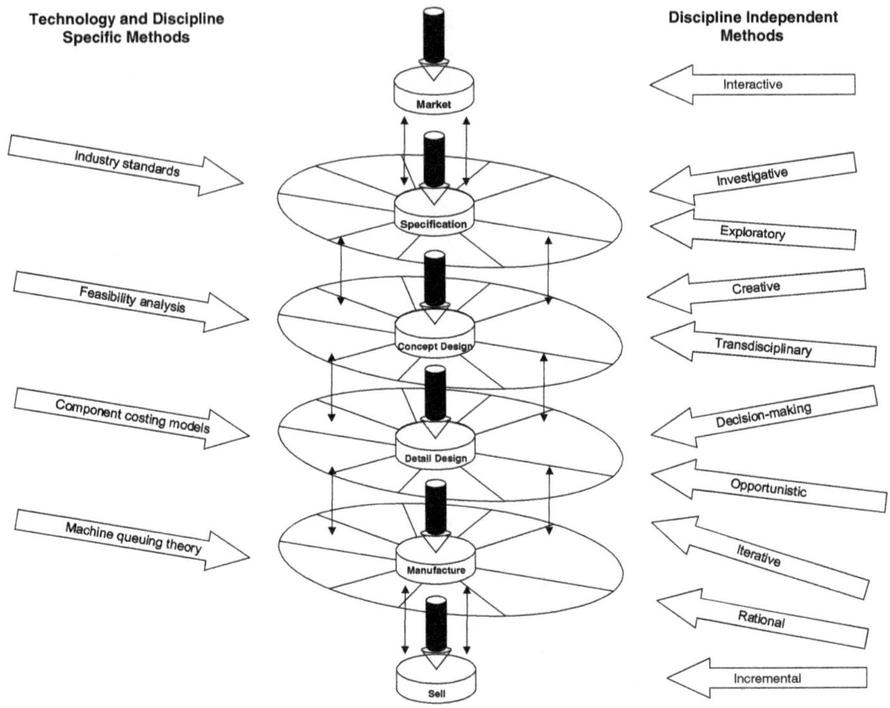

Fig. 2.11 Total design activity model

electromagnetic surveys, heat transfer studies, etc. The completed *Total Design Activity Model* is depicted in Fig. 2.11.

The *Total Design Activity Model* depicted in Fig. 2.11 includes examples of both technology and discipline specific methods and discipline independent methods to be illustrative of the inputs to the model. Real-world implementation of this model would involve many more methods. The details of this detailed model for to design are described in Pugh's seminal text *Total Design: Integrated Methods for Successful Product Engineering* (1991).

2.5.6 Methodology by the Association of German Engineers (VDI)

In Germany, the Association of German Engineers (VDI) has a formal guideline for the Systematic Approach to the Design of Technical Systems and Products (VDI 1987). The guideline proposes a generalized approach to the design of man-made systems that has wide applicability within a wide range of engineering disciplines. This approach is depicted in Fig. 2.12.

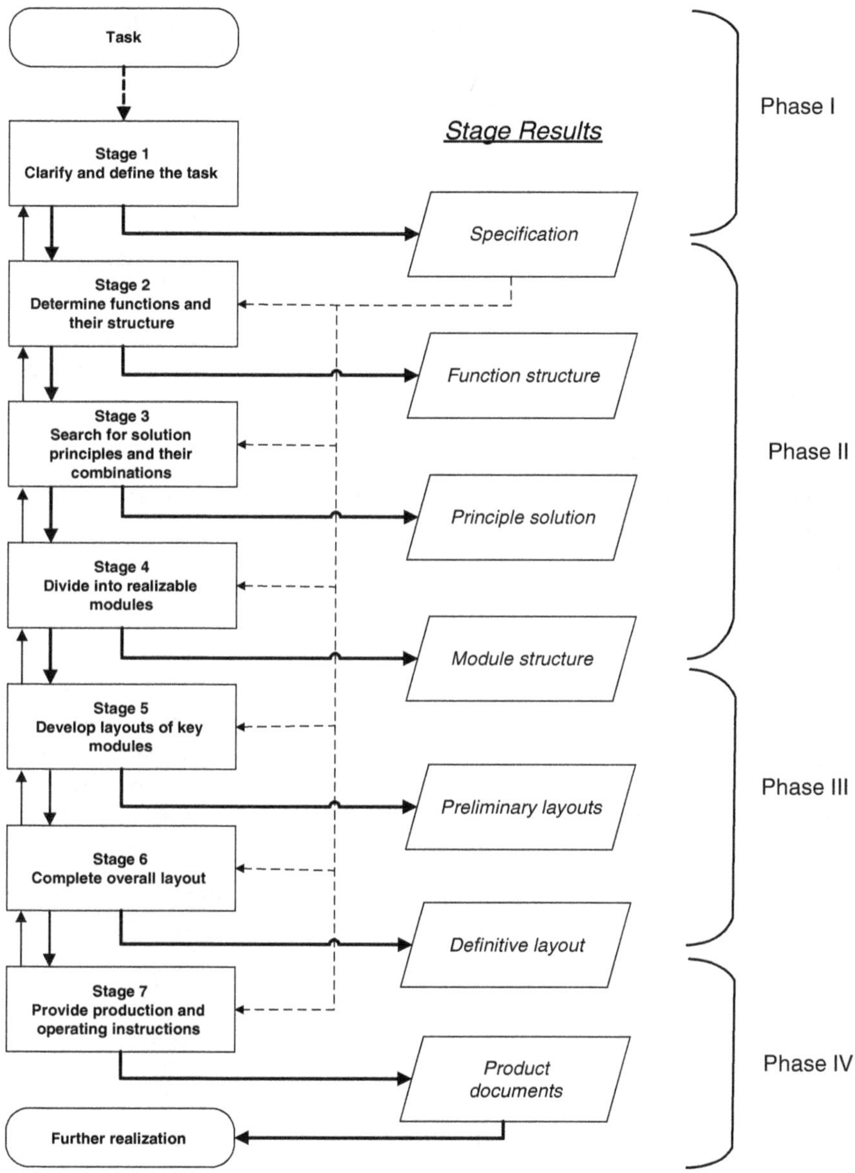

Fig. 2.12 General approach to design [adapted from Fig. 3.3 in (VDI 1987, p. 6)]

The model has four phases made up of seven stages and a specific result is associated with each stage. The approach in Fig. 2.12 should be "… regarded as a guideline to which detailed working procedures can be assigned. Special emphasis is placed on the iterative nature of the approach and the sequence of steps must not be considered rigid" (Pahl et al. 2011, p. 18).

2.5.7 Methodology by Pahl, Beitz, Feldhusen, and Grote

The team of Gerhard Pahl, Wolfgang Beitz, Jörg Feldhusen, and Karl-Heinrich Grote have authored one of the most popular textbooks on design, *Engineering Design: A Systematic Approach* (2011). In this text they propose of model for design that has four main phases: (1) planning and task clarification; (2) conceptual design; (3) embodiment design; and (4) detailed design. The simple nature of the model does not warrant a figure, but each of the phases are described in the following:

- Task Clarification—the purpose of this phase "is to collect information about the requirements that have to be fulfilled by the product, and also about the existing constraints and their importance" (Pahl et al. 2011, p. 131).
- Conceptual Design—the purpose of this phase is to determine the principle solution. "This is achieved by abstracting the essential problems, establishing function structures, searching for suitable working principles and then combining those principles into a working structure" (Pahl et al. 2011, p. 131).
- Embodiment Design—the purpose of this phase is to "determine the construction structure (overall layout) of a technical system in line with technical and economic criteria. Embodiment design results in the specification of a *layout*" (Pahl et al. 2011, p. 132).
- Detailed Design—the purpose of this phase is to finalize "the arrangement, forms, dimensions, and surface properties of all the individual parts are finally laid down, the materials specified, production possibilities assessed, costs estimated, and all the drawings and other production documents produced. The detailed design phase results in the *specification of information* in the form of *production documentation* (Pahl et al. 2011, p. 132).

The details of each of the phases in this model are presented in their text *Engineering Design: A Systematic Approach* (Pahl et al. 2011) which is now in its 3rd edition.

The section that follow will discuss an eighth design methodology—Axiomatic Design.

2.6 The Axiomatic Design Methodology

The *Axiomatic Design Methodology* is given special treatment in this chapter because it not only satisfies the Technical Processes in Table 2.2, but it also meets nine critical attributes that ensure a methodology remain sustainable (Adams and Keating 2011). The nine (9) critical attributes and how the Axiomatic Design Methodology satisfies these are presented in Table 2.8.

It is important to note that one of the most unique features of the *Axiomatic Design Methodology* (ADM) is its ability to not only satisfy the Technical

Table 2.8 Critical attributes of the axiomatic design methodology (ADM)

Critical Attribute	Attribute description and satisfaction in the ADM
1. Transportable	*A methodology must be capable of application across a spectrum of complex engineering problems and contexts.* The ADM has been successfully applied to a wide variety of design problems is multiple domains
2. Theoretical and Philosophical Grounding	*A valid methodology must have a linkage to a theoretical body of knowledge as well as philosophical underpinnings that form the basis for the methodology and its application.* The theoretical body of knowledge for the ADM is systems theory
3. Guide to Action	*A methodology must provide sufficient detail to frame appropriate actions and guide direction of efforts to implement the methodology.* The ADM provides clear guidance on how to transform customer requirements to functional and non-functional requirements to design parameters and process variables
4. Significance	*A methodology must exhibit the holistic capacity to address multiple problem domains, minimally including contextual, human, organizational, managerial, policy, technical, and political aspects.* The ADM addresses functional and non-functional requirements and systems constraints
5. Consistency	*A methodology must be capable of providing replicability of approach and results interpretation based on deployment of the methodology in similar contexts.* The ADM mathematical rigor ensures consistency of results
6. Adaptable	*A methodology must be capable of flexing and modifying the approach, configuration, execution, or expectations based on changing conditions or circumstances.* The ADM may be applied in a variety of conditions and circumstances subject to compliance with the axioms of systems theory
7. Neutrality	*A methodology attempts to minimize and account for external influences in application and interpretation.* The ADM is sufficiently transparency in technique to eliminate bias, surface assumptions, and account for limitations during execution of the methodology
8. Multiple Utility	*A methodology supports a variety of applications with respect to complex systems to include new system design, existing system transformation, and assessment of systems problems.* The ADM may be applied across multiple problem domains and applications
9. Rigorous	*A methodology must be capable of withstanding scrutiny with respect to: (1) identified linkage to a body of theory and knowledge; (2) sufficient depth to demonstrate detailed grounding in relationship to the theory and knowledge; and (3) capable of providing transparent, replicable results with accountability for explicit logic used to draw conclusions and interpretations.* The ADM's grounding in systems theory, application of axioms for information entropy and independence, and its mathematical rigor ensure replicable results that use common logic for the development of conclusions

Processes for design presented in Table 2.2, but its ability to invoke specific axioms of systems theory in order to develop quantitative measures for evaluating systems design endeavors. None of the seven design methodologies reviewed in Sect. 2.5 demonstrated that ability.

The sections that follow will introduce the basic elements of the ADM. The central focus will be on its ability to selection the best design alternative based upon a quantitative evaluation of the design's ability to satisfy its functional and non-functional requirements. The elimination of qualitative evaluation parameters and cost is a major shift from every other design methodology. As such, the Axiomatic Design Methodology is positioned as the premier methodology for systems design endeavors.

2.6.1 Introduction to the Axiomatic Design Methodology

The Axiomatic Design Methodology was developed by Professor Nam P. Suh while at the Massachusetts Institute of Technology. Professor Suh's design framework is founded upon two axioms of systems theory, that he titles the *independence axiom* and the *information axiom*. Suh uses these axioms, in conjunction with the concept of domains to develop a framework where customer attributes are transformed into process variables in a completed design. The basic idea of an axiomatic design framework was envisioned by Dr. Suh in the mid-1970s and was first published in 1990 (Suh 1990) and updated in 2001 (Suh 2001). The sections that follow will provide a high-level description of the Axiomatic Design Methodology.

2.6.2 Domains in the Axiomatic Design Methodology

A key concept in axiomatic design is that of domains. In the design world there are four domains: (1) the customer[1] domain, which is characterized by customer attributes that the customer and associated stakeholders would like to see in the their system; (2) The functional domain where the customer's detailed specifications, specified as both functional requirements (FR) and non-functional requirements (NFR) or what Suh describes as constraints (C) are specified; (3) The physical domain where the design parameters emerge; and (4) The process domain where process variables enable the design. Figure 2.13 is a depiction of the four domains of the design world.

[1]This chapter will adhere to Dr. Suh's term *customer*. However, note that this term is too narrowly focused. Therefore, the reader is encouraged to substitute the term stakeholder, which includes the larger super-set of those associated with any systems design.

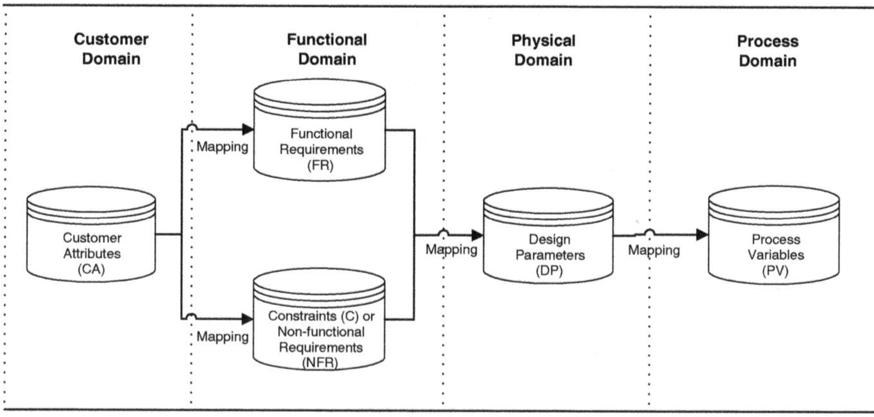

Fig. 2.13 Four domains of the design world

2.6.3 Independence Axiom

A second key concept of axiomatic design is the *independence axiom*. The independence axiom states:

> Maintain the independence of the functional requirements (Suh 2005b, p. 23).

Simply stated, each functional requirement[2] should be satisfied without affecting any other functional requirement. During the conceptualization process the functional requirements are transformed from the functional domain where they state *what,* to the physical domain where they will be met by *how*. The mapping should be one design parameter (DP) to one functional requirement (FR). Mathematically this can be related as two vectors, the FR vector $[FR]$ and the DP vector $[DP]$ as shown in Eq. 2.1.

Equation for Functional Requirements

$$[FR] = [A][DP] \tag{2.1}$$

where $[A]$ is the design matrix which relates FRs to DPs and is:

Equation for Design Matrix

$$[A] = \left\| \begin{matrix} A11 & A12 & A13 \\ A21 & A22 & A23 \\ A31 & A23 & A33 \end{matrix} \right\| \tag{2.2}$$

[2]Only functional requirements will be addressed in this description, but the concept also applies to the non-functional requirements that act as constraints on the system design.

Using the design matrix in Eqs. 2.2 and 2.1 may be written as Eq. 2.3.

Expanded Equation for Functional Requirements

$$\begin{aligned} FR_1 &= A_{11}DP_1 + A_{12}DP_2 + A_{13}DP_3 \\ FR_2 &= A_{21}DP_1 + A_{22}DP_2 + A_{23}DP_3 \\ FR_3 &= A_{31}DP_1 + A_{32}DP_2 + A_{33}DP_3 \end{aligned} \tag{2.3}$$

This satisfies the general relationship in Eq. 2.4.

General Equation for Functional Requirements

$$FR = \sum_{i=1}^{n} A_{ij}DP_j \tag{2.4}$$

where i = the number of Design Parameters (DP).

The independence axiom may be used to evaluate design complexity. Most systems exhibit complexity as a result of excessive interaction between components within the system design. Design complexity may be measured by observing system coupling, i.e., where the number of DPs is less than the number of FRs. In this case the design has added complexity because DPs are satisfying more than one FR, or the FR has not been satisfied.

The relevance of the independence axiom has additional utility in that individual designs may be evaluated, not qualitatively, but quantitatively, based on the relationship to an *ideal design*. The ideal design is one where the number of DPs are equal to the number of FRs, where the FRs are kept independent of one another. All design alternatives may be evaluated against the concept of an ideal design.

2.6.4 The Information Axiom

The information axiom is one of the seven axioms of systems theory (Adams et al. 2014). The Information Axiom states:

> Systems create, possess, transfer, and modify information. The information principles provide understanding of how information affects systems (Adams et al. 2014, p. 119).

The information axiom's principle invoked by Suh (1990, 2001) in his formulation for Axiomatic Design is the principle of information redundancy. Information redundancy is "the fraction of the structure of the message which is determined not by the free choice of the sender, but rather by the accepted statistical rules governing the use of the symbols in question" (Shannon and Weaver 1998, p. 13). It is the number of bits used to transmit a message minus the number of bits of actual information in the message.

The Axiomatic Design Methodology's information axiom makes use of Shannon's generalized formula for information entropy,[3] which is a measure of the uncertainty of the information, or the unpredictability of information content, as presented in Eq. 2.5.

Shannon's Equation for Information Entropy

$$H = -\sum p_i log p_i \tag{2.5}$$

where:

H information entropy

p_i probability of the information elements

The reformulated equation for information content (*I*), as related to the probability (p_i) of a design parameter (DP_i) satisfying a functional requirement (FR_i) is presented in Eq. 2.6.

System Information Content

$$I_{sys} = -\sum_{i=1}^{n} log_2 p_i \tag{2.6}$$

The information axiom, when used in this context, states that the system design with the smallest I_{sys} (i.e., the design with the least amount of information) is the best design. This is perfectly logical, because such a design requires the least amount of information to fulfill the design parameters.

The Axiomatic Design Methodology's utilization of Shannon's information entropy is remarkable because a system's design complexity, most often expressed as a qualitative assessment, may be represented as a quantitative measure based on the information entropy required to satisfy the design parameters.

2.6.5 Constraints or Non-functional Requirements

The design goals include not only the functional requirements (FR_i), but constraints (C_i) which place bounds on acceptable design solutions. Axiomatic design addresses two types of constraints: (1) input constraints, which are specific to the overall design goals and apply to all proposed designs; and (2) system constraints, which are specific to a particular system design.

---

[3]Information entropy is sometimes referred to as Shannon Entropy. For more information on Information Theory the reader may review either Ash (1965). *Information Theory*. New York: Dover Publications, or Pierce (1980). *An Introduction to Information Theory: Symbols, Signals and Noise* (2nd, Revised ed.). New York: Dover Publications.

> Constraints affect the design process by generating a specific set of functional requirements, guiding the selection of design solutions, and being referenced in design evaluation (Suh 2005a, p. 52).

The specific set of functional requirements generated by the constraints will be labeled *non-functional requirements*. Non-functional requirements (NFR) are addressed within the Axiomatic Design Methodology in the same manner as functional requirements. A discussion and nominal taxonomy for non-functional requirements will be presented in the next chapter.

2.7 Summary

In this chapter engineering design has been defined and positioned within the larger scientific paradigm and the engineering field. Desirable features and two modes of thought used in of engineering design were addressed. Terminology relating a methodology within a hierarchy of scientific approaches has been provided. Finally, seven historical and one preferred methodology for system design endeavors were presented.

The next chapter will review the definition for non-functional requirements and the role they play in every engineering design of man-made systems. It will also develop a notional taxonomy for identifying and addressing non-functional requirements in a system design endeavor.

References

Adams, K. M., Hester, P. T., Bradley, J. M., Meyers, T. J., & Keating, C. B. (2014). Systems theory: The foundation for understanding systems. *Systems Engineering, 17*(1), 112–123.

Adams, K. M., & Keating, C. B. (2011). Overview of the systems of systems engineering methodology. *International Journal of System of Systems Engineering, 2*(2/3), 112–119.

Angier, N. (2007). *The canon: A whirlwig tour of the beautiful basics of science*. New York: Houghton Mifflin Company.

Ash, R. B. (1965). *Information theory*. New York: Dover Publications.

Asimow, M. (1962). *Introduction to design*. Englewood Cliffs: Prentice-Hall.

BKCASE-Editorial-Board. (2014). *The guide to the systems engineering body of knowledge (SEBoK), version 1.3*. In R. D. Adcock (Ed.), Hoboken, NJ: The Trustees of the Stevens Institute of Technology.

Bourque, P., & Fairley, R. E. (Eds.). (2014). *Guide to the software engineering body of knowledge (version 3.0)*. Piscataway, NJ: Institute of Electrical and Electronics Engineers.

Carnap, R. (1934). On the character of philosophic problems. *Philosophy of Science, 1*(1), 5–19.

Checkland, P. B. (1999). *Systems thinking. Systems Practice*. Chichester: Wiley.

Cross, N. (2008). *Engineering design methods* (4th ed.). Hoboken, NJ: Wiley.

de Weck, O. L., Roos, D., & Magee, C. L. (2011). *Engineering systems: Meeting human needs in a complex technological world*. Cambridge, MA: MIT Press.

Eder, W. E. (2011). Engineering design science and theory of technical systems: Legacy of Vladimir Hubka. *Journal of Engineering Design, 22*(5), 361–385.

Evbuomwan, N. F. O., Sivaloganathan, S., & Jebb, A. (1996). A survey of design philosophies, models, methods and systems. *Proceedings of the Institution of Mechanical Engineers, Part B: Journal of Engineering Manufacture, 210*(4), 301–320.

French, M. J. (1998). *Conceptual design for engineers* (3rd ed.). London: Springer.

Honderich, T. (2005). *The Oxford companion to philosophy* (2nd ed.). New York: Oxford University Press.

Hubka, V., & Eder, W. E. (1995). *Design science: Introduction to the needs, scope and organization of engineering design knowledge* (2nd ed.). New York: Springer.

IEEE and ISO/IEC. (2008). IEEE and ISO/IEC Standard 15288: Systems and software engineering—system life cycle processes. New York: Institute of Electrical and Electronics Engineers and the International Organization for Standardization and the International Electrotechnical Commission.

IEEE and ISO/IEC. (2010). IEEE and ISO/IEC Standard 24765: Systems and software engineering—vocabulary. New York : Institute of Electrical and Electronics Engineers and the International Organization for Standardization and the International Electrotechnical Commission.

Kuhn, T. S. (1996). *The structure of scientific revolutions*. Chicago: University of Chicago Press.

Le Masson, P., Dorst, K., & Subrahmanian, E. (2013). Design theory: History, state of the art and advancements. *Research in Engineering Design, 24*(2), 97–103.

Mingers, J. (2003). A classification of the philosophical assumptions of management science methods. *Journal of the Operational Research Society, 54*(6), 559–570.

Mish, F. C. (Ed.). (2009). *Merriam-webster's collegiate dictionary* (11th ed.). Springfield, MA: Merriam-Webster, Incorporated.

Nagel, E. (1961). *The structure of science: Problems in the logic of scientific explanation*. New York: Harcourt, Brace & World.

Norman, D. A. (2013). *The design of everyday things* (Revised and expanded ed.). New York: Basic Books.

NSPE. (2013). *Engineering body of knowledge*. Washington, DC: National Society of Professional Engineers.

OECD. (2007). *Revised field of science and technology (FOS) classification in the frascati manual*. Paris: Organization for Economic Cooperation and Development.

Pahl, G., Beitz, W., Feldhusen, J., & Grote, K.-H. (2011). *Engineering design: A systematic approach* (K. Wallace & L. T. M. Blessing, Trans. 3rd ed.). Darmstadt: Springer.

Pierce, J. R. (1980). *An introduction to information theory: Symbols, signals & noise* (2nd Revised ed.). New York: Dover Publications.

Proudfoot, M., & Lacey, A. R. (2010). *The Routledge dictionary of philosophy* (4th ed.). Abingdon: Routledge.

Psillos, S. (2007). *Philosophy of science A-Z*. Edinburgh: Edinburgh University Press.

Pugh, S. (1991). *Total design: Integrated methods for successful product engineering*. New York: Addison-Wesley.

Runes, D. D. (Ed.). (1983). *The standard dictionary of philosophy*. New York: Philosophical Library.

Shannon, C. E., & Weaver, W. (1998). *The mathematical theory of communication*. Champaign, IL: University of Illinois Press.

Suh, N. P. (1990). *The principles of design*. New York: Oxford University Press.

Suh, N. P. (2001). *Axiomatic design: Advances and applications*. New York: Oxford University Press.

Suh, N. P. (2005a). Complexity in engineering. *CIRP Annals Manufacturing Technology, 54*(2), 46–63.

Suh, N. P. (2005b). *Complexity: Theory and applications*. New York: Oxford University Press.

VDI. (1987). *Systematic approach to the design of technical systems and products (VDI Guideline 2221)*. Berlin: The Association of German Engineers (VDI).

Chapter 3
Introduction to Non-functional Requirements

Abstract One of the most easily understood tasks during any systems design endeavor is to define the systems functional requirements. The functional requirements are a direct extension of the stakeholder's purpose for the systems and the goals and objectives that satisfy them. Less easily understood are a systems non-functional requirements, or the constraints under which the entire system must operate. Identification of non-functional requirements should happen early in the conceptual design stage of the systems life cycle, for the same reason that functional requirements are defined up-front—that is, costs sky-rocket when new requirements are added late in a systems design sequence. Approaches for addressing non-functional requirements are rarely addressed in texts on systems design. In order to provide a logical and repeatable technique for addressing over 200 existing non-functional requirements, they must be reduced parsimoniously to a manageable number. Over 200 non-functional requirements are reduced, using results reported in eight models from the extant literature. The 27 resultant non-functional requirements have been organized in a taxonomy that categorizes the 27 major non-functional requirements within four distinct categories. Utilization of this taxonomy provides a framework for addressing non-functional requirements during the early system design stages.

3.1 Introduction to Non-functional Requirements

This chapter will begin by reviewing the definition for non-functional requirements and the roles they play in the engineering design of man-made systems. The chapter will then address the wide range of non-functional requirements and introduce a number of taxonomies used to describe non-functional requirements. It will conclude by presenting a notional taxonomy or framework for understanding non-functional requirements and their role as part of any system design endeavor.

The chapter has a specific learning goal and associated objectives. The learning goal of this chapter is to be able to describe nonfunctional requirements and a

K.M. Adams, *Non-functional Requirements in Systems Analysis and Design*,
Topics in Safety, Risk, Reliability and Quality 28,
DOI 10.1007/978-3-319-18344-2_3

taxonomy for addressing them during systems design endeavors. This chapter's goal is supported by the following objectives:

- Define a non-functional requirement.
- Discuss three aspects of non-functional requirements that complicate dealing with them.
- Name 10 non-functional requirements.
- Discuss the historical development of frameworks for non-functional requirements.
- Describe the elements of the Taxonomy of NFR for Systems.
- Describe the four-level structural map for measuring the attributes of non-functional requirements.

3.2 Definitions for Functional and Non-functional Requirements

All system design efforts start with a concept for the system. Concepts start as ideas and are moved forward is a series of actions during the concept design stage of the systems life cycle. As the system life cycle progresses the system's concept is transposed into formal system-level requirements where the stakeholder's needs are transformed into discrete requirements.

During the concept design stage two requirements-related technical processes, depicted in Table 3.1, are invoked.

The requirements addressed in Table 3.1 are functional requirements.

3.2.1 Functional Requirements

A functional requirement, as defined in the standard for systems and software engineering vocabulary is defined as:

Table 3.1 Requirements-related technical processes in the concept design stage

Technical process	Purpose
Stakeholder requirements definition	"Define the requirements for a system that can provide the services needed by users and other stakeholders in a defined environment" IEEE and ISO/IEC (2008, p. 36)
Requirements analysis	"Transform the stakeholder, requirement-driven view of desired services into a technical view of a required product that could deliver those services" IEEE and ISO/IEC (2008, p. 39)

Table 3.2 Definitions for functional requirement

Definition	Source
"These refer largely to what the system should do. These requirements should be action oriented and should describe the tasks or activities that the system performs during its operation"	Kossiakoff et al. (2011, p. 145)
"Functional requirements relate to specific functions (at any level of abstraction) that the system must perform while transforming inputs into outputs. As a result, a functional requirement is a requirement that can be associated with one or more of the system's outputs"	Buede (2000, p. 130)

> 1. A statement that identifies what a product or process must accomplish to produce required behavior and/or results. 2. A requirement that specifies a function that a system or system component must be able to perform. (IEEE and ISO/IEC 2010, p. 153)

There are additional definitions which will provide additional insight about this category of requirement. Table 3.2 provides definitions from two of the most popular systems engineering and systems design texts.

From these two definitions it is clear that functional requirements have the following essential characteristics (note that all of these are characterized by verbs):

1. Define *what* the system should do.
2. Be action oriented.
3. Describe tasks or activities.
4. Are associated with the transformation of inputs to outputs.

These are the requirements that the Axiomatic Design Methodology describes as FR_i which are mapped to Design Parameters DP_i during the transformation from the functional domain to the physical domain as part of the Axiomatic Design Methodology described in Chap. 2.

In the Axiomatic Design Methodology the design goals include not only the functional requirements (FR_i), but constraints (C_i) which place bounds on acceptable design solutions. Axiomatic design addresses two types of constraints: (1) input constraints, which are specific to the overall design goals and apply to all proposed designs; and (2) system constraints, which are specific to a particular system design.

> Constraints affect the design process by generating a specific set of functional requirements, guiding the selection of design solutions, and being referenced in design evaluation. (Suh 2005, p. 52)

The specific set of functional requirements generated by the constraints will be labeled *non-functional requirements.* Non-functional requirements (NFR) are addressed within the Axiomatic Design Methodology in the same manner as functional requirements. A discussion of non-functional requirements will be presented in the next section.

3.2.2 Non-functional Requirements

Non-functional requirements, as defined in the standard for systems and software engineering vocabulary are defined as:

> A software requirement that describes not what the software will do but how the software will do it. Syn: design constraints, non-functional requirement. EXAMPLE software performance requirements, software external interface requirements, software design constraints, and software quality attributes. Nonfunctional requirements are sometimes difficult to test, so they are usually evaluated subjectively. (IEEE and ISO/IEC 2010, p. 231)

While the *IEEE Guide for Developing System Requirements Specifications* (IEEE 1998b) is silent on non-functional requirements, the *IEEE Recommended Practice for Software Requirements Specifications* (IEEE 1998a) states that requirements consist of "functionality, performance, design constraints, attributes, or external interfaces" (p. 5). Descriptions of what each software requirement is supposed to answer are presented in Table 3.3.

There are additional definitions which will help give additional insight about non-functional requirements. Table 3.4 provides definitions from a variety of systems and software engineering and design texts.

There are three additional aspects to non-functional requirements that complicate the situation.

1. Non-functional requirements can be '**subjective**', since they can be viewed, interpreted and evaluated differently by different people. Since NFRs are often stated briefly and vaguely, this problem is compounded. (Chung et al. 2000, p. 6)
2. Non-functional requirements can also be '**relative**', since the interpretation and importance of NFRs may vary on the particular system being considered. Achievement of NFRs can also be relative, since we may be able to improve upon existing ways to achieve them. For these reasons a 'one solution fits all' approach may not be suitable. (Chung et al. 2000, pp. 6–7)

Table 3.3 Requirements and questions they should answer

Requirement	Questions requirements should answer
Functionality	What is the software supposed to do?
Performance	What is the speed, availability, response time, recovery time of various software functions, etc.?
Design constraints	Are there any required standards in effect, implementation language, policies for database integrity, resource limits, operating environment(s) etc.?
Attributes	What are the portability, correctness, maintainability, security, etc. considerations?
External interfaces	How does the software interact with people, the systems hardware, other hardware, and other software?

Table 3.4 Definitions for non-functional requirement

Definition	Source
"A requirement that describes not what the software will do, but how the software will do it is called a nonfunctional requirement (NFR)"	Ebert (1998, p. 175)
"Describes a restriction on the system that limits our choices for constructing a solution to the problem"	Pfleeger (1998, p. 141)
"A description of a property or characteristic that a software system must exhibit or a constraint that it must respect, other than an observable system behavior"	Wiegers (2003, p. 486)
"NFRs state constraints to the system as well as particular notions of qualities a system might have, for example, accuracy, usability, safety, performance , reliability, security. NFRs constrain 'how' the system must accomplish 'what'"	Cysneiros and Yu (2004, p. 116)
"Properties or qualities the product must have to facilitate its functionality"	Robertson and Robertson (2005, p. 146)
"These are constraints on the services or functions offered by the system. They include timing constraints, constraints on the development process and standards. Non-functional requirements often apply to the system as a whole. They do not usually just apply to individual system features or services"	Somerville (2007, p. 119)

3. Furthermore, non-functional requirements can often be '**interacting**', in that attempts to achieve one NFR can hurt or help the achievement of other NFRs. SAs NFRs have a global impact on systems, localized solutions may not suffice. (Chung et al. 2000, p. 7)

The third aspect of NFRs, *interacting*, is an extremely important one.

> One important step during specification of NFR is conflict resolution of interacting NFRs. This is more relevant in the case of NFRs compared to functional requirements as there are in most systems severe trade-offs among the major groups of NFRs. One might go even as far as stating that there are inherent contradictions among distinct NFRs. For example, performance often interferes with maintainability and reusability. (Ebert 1998, p. 178)

From these definitions and conditions practitioners may easily conclude that non-functional requirements have the following essential characteristics (note that all of these are characterized by adjectives that define *which*, *what kind of*, or *how many*):

1. Define a property or quality that the system should have.
2. Can be subjective, relative, and interacting.
3. Describe how well the systems must operate.
4. Are associated with the entire system.

Practitioners responsible for systems design could benefit from a structured approach to the identification, organization, analysis, and refinement of non-functional requirements in support of design activities (Cleland-Huang et al. 2007;

Cysneiros and do Prado Leite 2004; Gregoriades and Sutcliffe 2005; Gross and Yu 2001; Sun and Park 2014).

It is precisely because non-functional requirements (NFR) describe important, and very often critical requirements, that a formal, structured approach for their identification, organization, analysis, and refinement is required as a distinct element of systems design. Non-functional requirements include a broad range of system needs that play a critical role in early development of the systems architecture (Nuseibeh 2001). Failure to formally identify and account for non-functional requirements early in a system design may prove to be costly in later stages of the systems life cycle. In fact, "failing to meet a non-functional requirements can mean that the whole system is unusable" (Somerville 2007, p. 122).

3.2.3 A Structure for Non-functional Requirements

The current state of affairs with respect to non-functional requirements has shown that:

- There is not a single, agreed upon, formal definition
- There is not a complete list.
- There is not a single universal classification schema, framework, or taxonomy.

The three sections that follow will: (1) Present a list of non-functional requirements with appropriate formal definitions; (2) Review the historical work and research associated with the development of a universal classification schema, framework or taxonomy for non-functional requirements; and (3) Recommend a notional model for understanding the major non-functional requirements in systems design.

3.3 Identification and Organization of Non-functional Requirements

This section will identify and define the principal non-functional requirements associated with systems. It is important to recognize that non-functional requirements span the complete life cycle of a system, from conception to retirement and disposal and that each non-functional requirement has its *20 min of fame* and is accompanied by its own set of experts, zealots, and benefactors.

In their seminal work *Engineering Systems: Meeting Human Needs in a Complex Technological World*, de Weck et al. (2011) of the Massachusetts Institute of Technology, discuss the relationship between non-functional requirements and what they term *ilities*.

> In computer science ilities are discussed as nonfunctional requirements. (de Weck et al. 2011, p. 196)
> Ilities are requirements of systems, such as flexibility or maintainability, often ending in the suffix "ility"; properties of systems that are not necessarily part of the fundamental set of functions or constraints and sometimes not in the requirements. (de Weck et al. 2011, p. 187)

In their chapter on life-cycle properties of systems de Weck et al. (2011) attribute the increase in *nonfunctional requirements*, *life-cycle properties*, or *ilities* to the complexity of modern systems and the scale of their deployments and the important side effects of their ubiquitous presence in the modern era. They provide an excellent discussion of the history associated with the expansion of the four classic systems *ilities*—safety, quality, usability, and reliability—to the plethora of *ilities* present in systems endeavors today. In fact, there are over 200 recognized *ilities* in use in systems endeavors. Table 3.5 is an alphabetical list of 161 non-functional requirements (Chung et al. 2000, p. 160) used in software systems engineering. Table 3.6 includes an additional 38 non-functional requirements from the extant literature (Mairiza et al. 2010, p. 313) that were not included in Table 3.5. Finally, Table 3.7 is a list of 19 non-functional requirements from non-attributable sources.

A review of the 218 non-functional requirements just presented includes just about any *-ility* that can be imagined. Treatment of each of the 218 non-functional requirements presented in Tables 3.5, 3.6, and 3.7 is not practical in a single presentation. However, practitioners should be aware that the list of non-functional requirements continues to wax and wane as new life cycle properties are required to evaluate, constrain, and measure systems. It is important to note that although not specifically included in many requirements documents, each of the over 200 non-functional requirements often play a critical role during systems design activities. In most cases the non-functional requirements are used as selection criteria when choosing from an array of design alternatives.

The section that follows will review the historical work and research associated with the development of a universal classification schema, framework or taxonomy for non-functional requirements.

3.4 Classification Models for Non-functional Requirements

The quest to identify and organize non-functional requirements started in 1976 and continues to this day. This section will review some of the major classification models for non-functional requirements.

Table 3.5 A list of non-functional requirements (Chung et al. 2000, p. 160)

Accessibility	Degradation of service	Modularity	Security
Accountability	Dependability	Naturalness	Sensitivity
Accuracy	Development cost	Nomadicity	Similarity
Adaptability	Development time	Observability	Simplicity
Additivity	Distributivity	Off-peak period performance	Software cost
Adjustability	Diversity	Operability	Software production time
Affordability	Domain analysis cost	Operating cost	Space boundedness
Agility	Domain analysis time	Peak-period performance	Space performance
Auditability	Efficiency	Performability	Specificity
Availability	Elasticity	Performance	Stability
Buffer space performance	Enhanceability	Planning cost	Standardizability
Capability	Evolvability	Planning time	Subjectivity
Capacity	Execution cost	Plasticity	Supportability
Clarity	Extensibility	Portability	Surety
Code-space performance	External consistency	Precision	Survivability
Cohesiveness	Fault-tolerance	Predictability	Susceptibility
Commonality	Feasibility	Process management time	Sustainability
Communication cost	Flexibility	Productivity	Testability
Communication time	Formality	Project stability	Testing time
Compatibility	Generality	Project tracking cost	Throughput
Completeness	Guidance	Promptness	Time performance
Component integration cost	Hardware cost	Prototyping cost	Timeliness
Component integration time	Impact analyzability	Prototyping time	Tolerance
Composability	Independence	Reconfigurability	Traceability
Comprehensibility	Informativeness	Recoverability	Trainability
Conceptuality	Inspection cost	Recovery	Transferability
Conciseness	Inspection time	Reengineering cost	Transparency
Confidentiality	Integrity	Reliability	Understandability
Configurability	Inter-operability	Repeatability	Uniform performance
Consistency	Internal consistency	Replaceability	Uniformity
Controllability	Intuitiveness	Replicability	Usability
Coordination cost	Learnability	Response time	User-friendliness
Coordination time		Responsiveness	Validity

(continued)

Table 3.5 (continued)

	Main-memory performance		
Correctness	Maintainability	Retirement cost	Variability
Cost	Maintenance cost	Reusability	Verifiability
Coupling	Maintenance time	Risk analysis cost	Versatility
Customer evaluation time	Maturity	Risk analysis time	Visibility
Customer loyalty	Mean performance	Robustness	Wrappability
Customizability	Measurability	Safety	
Data-space performance	Mobility	Scalability	
Decomposability	Modifiability	Secondary-storage performance	

Table 3.6 Additional non-functional requirements (Mairiza et al. 2010, p. 313)

Analyzability	Demonstrability	Manageability
Anonymity	Durability	Performance
Atomicity	Effectiveness	Privacy
Attractiveness	Expandability	Provability
Augmentability	Expressiveness	Quality of service
Certainty	Extendability	Readability
Changeability	Functionality	Self-descriptiveness
Communicativeness	Immunity	Structuredness
Complexity	Installability	Suitability
Comprehensiveness	Integratability	Tailorability
Conformance	Legibility	Trustability
Debuggability	Likeability	Viability
Defensibility	Localizability	

Table 3.7 Non-attributed non-functional requirements

Degradability	Heterogeneity	Reproducibility
Deployability	Homogeneity	Resilience
Determinability	Interchangeability	Securability
Disposability	Manufacturability	Serviceability
Distributability	Producability	Ubiquity
Expandability	Repairability	
Fidelity	Repeatability	

3.4.1 Boehm's Software Quality Initiative

Barry Boehm and two of his colleague at TRW conducted a study (Boehm et al. 1976) which produced 23 non-functional characteristics of software quality that they arranged in a hierarchical tree. The lower-level branches in the tree contain sub-characteristics of the higher-level characteristic. In the schema presented in Fig. 3.1 the lower-level characteristics in the tree are necessary but not sufficient for achieving the higher-level characteristics.

3.4.2 Rome Air Development Center Quality Models

A number of models were developed at the United States Air Force's Rome Air Development Center between 1978 and 1985. Three of these models are presented in the following sections.

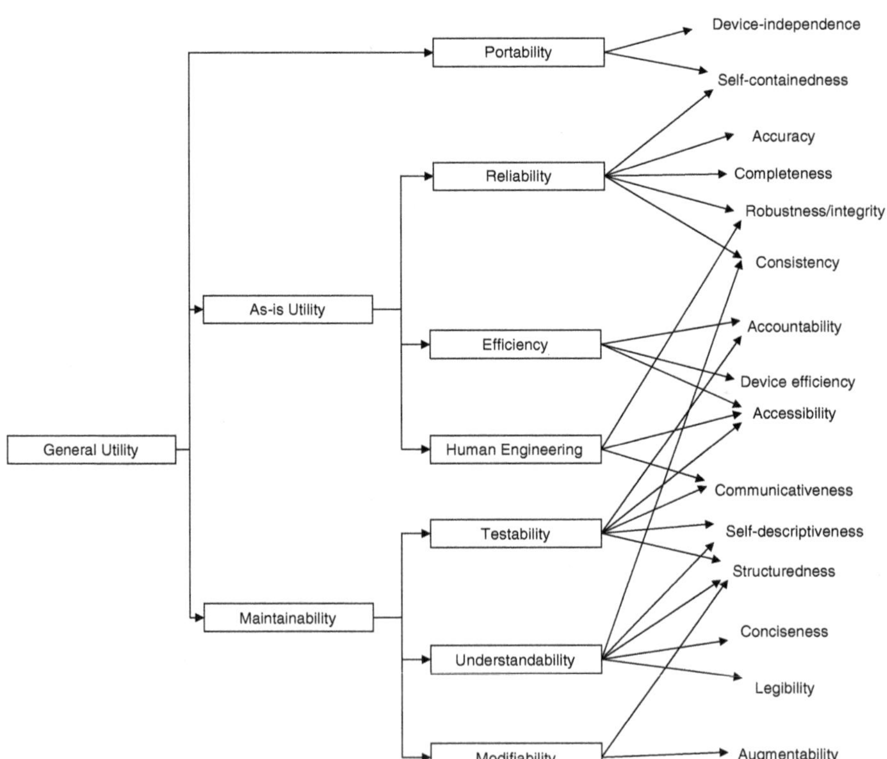

Fig. 3.1 Software quality characteristics tree [adapted from Fig. 1 in (Boehm et al. 1976, p. 595)]

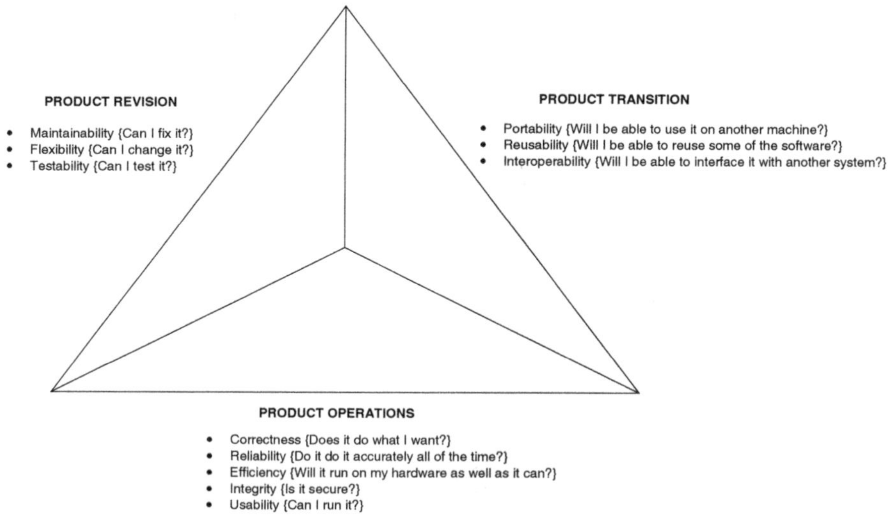

Fig. 3.2 Software quality factors [adapted from Fig. 2 in (Cavano and McCall 1978, p. 136)]

3.4.2.1 Cavano and McCall's Model

Cavano and McCall (1978) conducted a study which organized 11 non-functional quality factors by the lifecycle phase in which they were deemed to be most important for a developed system. The main goal of the study was to bridge the gap between users and developers and the ensuing model mapped the user's view with the priority of the software system development organization. The three development organization's perspectives and principal questions were: (1) Product revision—how easy is it to correct errors and add revisions? (2) Product transition—how easy is it to adapt to changes in the technical environment? (3) Product operation—how well does the system operate? The three perspectives and 11 non-functional quality factors with associated questions are depicted in Fig. 3.2.

3.4.2.2 McCall's and Masumoto's Factor Model Tree

McCall and his colleague Mike Masumoto, under the direction of James P. Cavano, continued the work in non-functional quality requirements and developed the *Software Quality Measurement Manual* (McCall and Matsumoto 1980). Their Quality-Factor tree is depicted in Fig. 3.3.

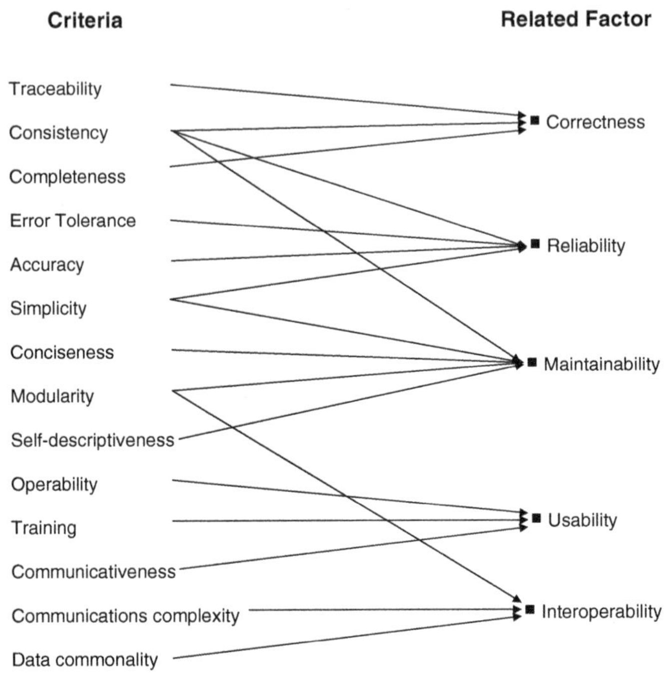

Fig. 3.3 USAF quality-factor tree [adapted from (McCall and Matsumoto 1980, p. 24)]

3.4.2.3 Software Quality Evaluation Guide

The final work on software quality was conducted between 1982 and 1984 and resulted in the third volume of the Software Quality Evaluation Guidebook (Bowen et al. 1985). The guidebook provides a comprehensive set of procedures and techniques to enable data collection personnel to apply quality metrics to software products and to evaluate the achieved quality levels. The associated model had 3 acquisition concerns, 13 quality factors, 29 criteria, 73 metrics, and over 300 metric elements. Table 3.8 shows the relationship between the acquisition concern, quality factors and criteria.

3.4.3 FURPS and FURPS+ Models

The FURPS Model was first introduced by Robert Grady and Deborah Caswell (Grady and Caswell 1987). The model's acronym is based on its five categories: (1) functionality; (2) usability; (3) reliability; (4) performance; and (5) supportability. The original FURPS Model "was extended to empathize various specific attributes" (Grady 1992, p. 32) and re-designated FURPS+. The FURPS+ categories and

Table 3.8 Software quality evaluation guidebook model (Bowen et al. 1985)

System need factor and acquisition concern	Quality factor and user concern	Criteria
Performance factor attributes —*How well does it function?*	Efficiency—*How well does it utilize a resource?*	Effectiveness—communication
		Effectiveness—processing
		Effectiveness—storage
	Integrity—*How secure is it?*	System accessibility
	Reliability—*What confidence can be placed in what it does?*	Accuracy
		Anomaly management
		Simplicity
	Survivability—*How well will it perform under adverse conditions?*	Anomaly management
		Autonomy
		Distributedness
		Modularity
		Reconfigurability
	Usability—*How easy it is to use?*	Operability
		Training
Design factor attributes—*How valid is the design?*	Correctness—*How well does it conform to the requirements?*	Completeness
		Consistency
		Traceability
	Maintainability—*How easy is it to repair?*	Consistency
		Modularity
		Self-descriptiveness
		Simplicity
		Visibility
	Verifiability—*How easy is it to verify its performance?*	Modularity
		Self-descriptiveness
		Simplicity
		Visibility
Adaptation factor attributes —*How adaptable is it?*	Expandability—*How easy is it to expand or upgrade its capability or performance?*	Augmentability
		Generality
		Modularity
		Self-descriptiveness
		Simplicity
		Virtuality

(continued)

Table 3.8 (continued)

System need factor and acquisition concern	Quality factor and user concern	Criteria
	Flexibility—*How easy is it to change?*	Generality
		Modularity
		Self-descriptiveness
		Simplicity
	Interoperability—*How easy is it to interface with another system?*	Commonality
		Functional overlap
		Independence
		Modularity
		System compatibility
	Portability—*How easy is it to transport?*	Independence
		Modularity
		Self-descriptiveness
	Reusability—*How easy is it to convert for use in another application?*	Application independence
		Document accessibility
		Functional scope
		Generality
		Independence
		Modularity
		Self-descriptiveness
		Simplicity
		System clarity

attributes are depicted in Table 3.9 (Grady 1992, p. 32). The FURPS+ elements represent a number of the non-functional requirements presented in Tables 3.5, 3.6 and 3.7.

3.4.4 Blundell, Hines and Stach's Quality Measures

James K. Blundell, Mary Lou Hines, and Jerrold Stach of the University of Missouri—Kansas City (Blundell et al. 1997) developed a highly detailed non-functional quality measurement model that includes 39 quality measures that are

Table 3.9 FURPS model categories and attributes (Grady 1992, p. 32)

Category	Attribute
Functionality	Feature set
	Capabilities
	Generality
	Security
Usability	Human factors
	Aesthetics
	Consistency
	Documentation
Reliability	Frequency/severity of failure
	Recoverability
	Predictability
	Accuracy
	Mean time to failure
Performance	Speed
	Efficiency
	Resource consumption
	Thruput [sic]
	Response time
Supportability	Testability
	Extensibility
	Adaptability
	Maintainability
	Compatibility
	Configurability
	Serviceability
	Installability
	Localizability

related to 18 characteristics, which are then each related to seven critical design attributes. The seven critical design attributes are shown in Table 3.10.
The critical design characteristics are related to 18 characteristics desirable in a software system. This relationship is shown in Table 3.11.
The final relationship is between the 18 desired characteristics and the 39 measures of quality, or *ilities*, which is related in Table 3.12.
The most intriguing feature of this model is relationship between the 39 non-functional quality measures and the seven design attributes (cohesion, complexity, coupling, data structure, intra-modular complexity, inter-modular complexity, and token selection). The least appealing feature of the model is that the 39 non-functional quality measures are neither organized nor related, leaving the user to face a huge array of relationships.

Table 3.10 Critical design attributes (Blundell et al. 1997, pp. 244–245)

Design attribute	Attribute description
Cohesion (COH)	The singularity of function of a single module
Complexity (COM)	The complexity within modules
Coupling (COU)	The simplicity of the connection between modules
Data structures (DAS)	Data types based upon functional requirements
Intra-modular complexity (ITA)	The complexity within modules
Inter-modular complexity (ITE)	The complexity between modules
Token selection (TOK)	The number of distinct lexical tokens in the program code

Table 3.11 Relationship between design attributes and desired characteristics (Blundell et al. 1997, p. 343)

#	Characteristic	COH	COM	COU	DAS	ITA	ITE	TOK
1	Conciseness		✓	✓	✓	✓	✓	✓
2	Ease of change	✓	✓	✓	✓		✓	✓
3	Ease of checking conformance		✓			✓	✓	
4	Ease of coupling to other systems		✓	✓			✓	
5	Ease of introduction of new features	✓	✓	✓			✓	
6	Ease of testing	✓	✓	✓	✓	✓	✓	✓
7	Ease of understanding	✓	✓	✓	✓	✓	✓	✓
8	Freedom from error	✓	✓		✓	✓	✓	✓
9	Functional independence of modules	✓		✓		✓	✓	
10	Precise computations				✓	✓		✓
11	Precise control	✓			✓	✓		✓
12	Shortest loops		✓		✓	✓		✓
13	Simplest arithmetic operators		✓			✓		✓
14	Simplest data types		✓		✓	✓	✓	
15	Simplest logic		✓		✓	✓		✓
16	Standard data types				✓	✓		
17	Ease of maintenance	✓	✓	✓	✓	✓	✓	✓
18	Functional specification compliance		✓	✓				

Table 3.12 Relationship between desired characteristics and quality measures (Blundell et al. 1997, pp. 236–237)

#	Measure of quality	Associated desired characteristics
1	Accuracy	8, 10, 11
2	Adaptability	5
3	Auditability	3
4	Availability	None listed
5	Chang[e]ability	2
6	Completeness	18
7	Conciseness	10
8	Consistency	None listed
9	Correctness	8, 18
10	Data commonality	14, 16
11	Dependability	=reliability, 18
12	Efficiency	4, 9, 12, 13, 14, 15, 16
13	Error tolerance	None listed
14	Expandability	2
15	Flexibility	2
16	Functionality	18
17	Generality	4, 9
18	Hardware independence	None listed
19	Human factors	None listed
20	Integrity	None listed
21	Interoperability	4
22	Maintainability	2, 6, 7, 17
23	Modifiability	2
24	Modularity	9
25	Operability	7
26	Portability	17
27	Reliability	18
28	Reusability	4, 7, 9
29	Robustness	18
30	Security	None listed
31	Self documentation	7
32	Simplicity	7
33	Supportability	2
34	Testability	6
35	Traceability	7
36	Transportability	=portability, 17
37	Understandability	7
38	Usability	None listed
39	Utility	None listed

3.4.5 Somerville's Classification Schema

Ian Somerville, the author of a major text on software engineering (Somerville 2007), has offered a classification schema for non-functional requirements that places them in three general groups: (1) process considerations; (2) product considerations; and (3) external considerations. This model is depicted in Fig. 3.4.

Somerville's schema offers no insight into its development but provides the following guidelines for its three top-levels.

- Product requirements. *These requirements specify product behavior.*
- Organizational requirements. *These requirements are derived from policies and procedures in the customer's and developer's organisations* [sic].
- External Requirements. *This broad heading covers all requirements that are derived from factors external to the system and its development process.* (Somerville 2007, p. 123)

Somerville states that non-functional requirements can be difficult to verify and "whenever possible, you should write requirements quantitatively so that they can be objectively tested" (Somerville 2007, p. 124).

3.4.6 International Standards

Non-functional requirements have been addressed in international standards as part of the software and systems quality initiative. Both the earlier ISO/IEC Standard

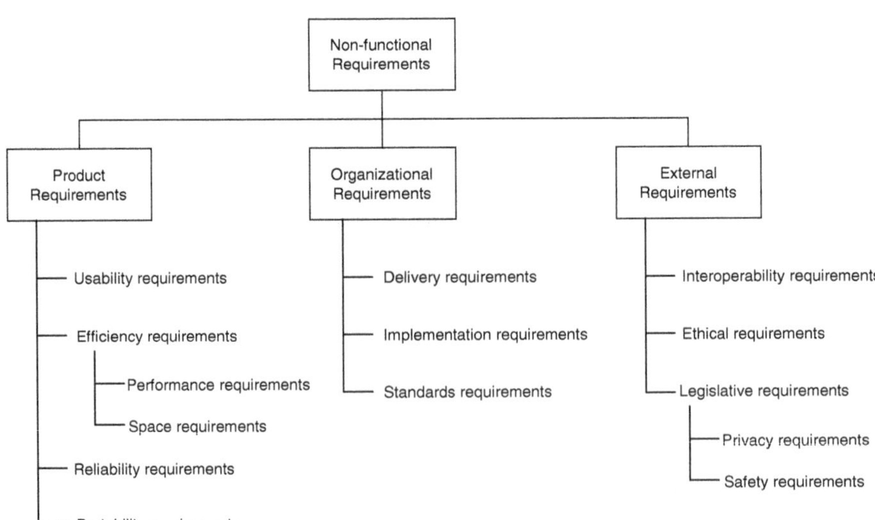

Fig. 3.4 Somerville's NFR classification schema (Somerville 2007, p. 122)

9126 (ISO/IEC 1991) and its replacement ISO/IEC Std 25010 (ISO/IEC 2011) include non-functional requirements, definitions, and how to measure them as part of a systems endeavor. Table 3.13 lists the non-functional requirements addressed in the latest international standard for systems quality, ISO/IEC Standard 25010: Systems and software engineering—Systems and software Quality Requirements and Evaluation (SQuaRE)—System and software quality models. The standard has eight main characteristics, each with a set of supporting sub-characteristics.

3.4.7 NFR Elicitation

Understanding non-functional requirements (NFR) requires the design team to understand the system's domain as part of the formal elicitation process. A formal process for elicitation of non-functional requirements is provided in two sources: (1) *Non-functional Requirements in Software Engineering* (Chung et al. 2000) and (2) as a chapter in *Perspectives on Software Requirements* (Cysneiros and Yu 2004). Although they have titles with the word software, both of these sources provide detailed techniques for understanding and dealing with non-functional requirements as part of the larger systems design process.

The section that follows will recommend a notional model for understanding the major non-functional requirements in systems design.

3.5 Notional Framework for Understanding Major NFR in Systems Design

In this section a notional model for understanding the major non-functional requirements in systems design, principally based upon the historical work presented in the previous section, will be developed. Because treatment of each of the non-functional requirements previously presented is not practical in a single presentation, they will be reduced to a manageable number that adequately represents the major non-functional requirements that will be required to be addressed in all major systems design endeavors.

3.5.1 Rationalization of Non-functional Requirements Classification Schemas

The first step is to rationalize the classification schemas for non-functional requirements from the extant literature. Table 3.14 relates each of the eight classification schemas (or models) by comparing the number of categories or factors

Table 3.13 Non-functional requirements in ISO/IEC Std 25010

Characteristic	Sub-characteristics
1. *Functional suitability*—The degree to which the product provides functions that meet stated and implied needs when the product is used under specified conditions	• Appropriateness • Completeness • Correctness
2. *Reliability*—The degree to which a system or component performs specified functions under specified conditions for a specified period of time	• Availability • Fault Tolerance • Recoverability • Maturity
3. *Usability—The degree to which the product has attributes that enable it to be understood, learned, used and attractive to the user, when used under specified conditions*	• Accessibility • Appropriateness • Learnability • Operability • User error protection • User interface aesthetics
4. *Performance efficiency*—The performance relative to the amount of resources used under stated conditions	• Time Behaviour • Resource Utilization
5. *Security*—The degree of protection of information and data so that unauthorized persons or systems cannot read or modify them and authorized persons or systems are not denied access to them	• Confidentiality • Integrity • Non-repudiation • Accountability • Authenticity
6. *Compatibility*—The degree to which two or more systems or components can exchange information and/or perform their required functions while sharing the same hardware or software environment	• Co-existence • Interoperability
7. *Maintainability*—The degree of effectiveness and efficiency with which the product can be modified	• Analyzability • Modifiability • Modularity • Reusability • Testability
8. *Portability*—The degree to which a system or component can be effectively and efficiently transferred from one hardware, software or other operational or usage environment to another	• Adaptability • Installability • Replaceability

and characteristics and lists the number of unique categories, factors, or criteria as non-functional requirements to be considered for inclusion in the notional framework. In this fashion the large body of non-functional requirements are reduced to 209.

By considering only the unique categories, factors, or criteria between the eight models the total number of non-functional requirements treated in the extant literature is reduced from 209 to 96.

Table 3.14 Functional requirements and major classification models

Model	Category or factors	Criteria	Total
Boehm et al. (1976)	8	16	24
Cavano and McCall (1978)	–	11	11
McCall and Masumoto (1980)	5	14	19
Bowen et al. (1985)	13	23	36
Grady and Caswell (1987); Grady (1992)	4	19	23
Blundell et al. (1997)	6	38	44
Somerville (2007)		14	14
ISO/IEC Std 25010 (2011)	8	30	38
Total	44	165	209
Unique items			96

3.5.2 Unique Non-functional Requirements

An analysis of the criteria in each of the seven historical models shows that not all of the criteria are universally applied. Table 3.15 reveals the frequency of criteria in the eight models.

The decision about which of the criteria in Table 3.15 to consider for inclusion in the notional framework is aided by one final task, reviewing the established formal definitions for each of the 96 non-functional requirements criteria.

3.5.3 Formal Definitions for Most Frequent Non-functional Requirements

Table 3.16 provides an alphabetical list and the formal definitions from *IEEE Standard 24765, Systems and Software Engineering—Vocabulary*[1] for 24 non-functional requirements criteria that achieved a frequency of 3 or higher. Table 3.16 also includes definitions for three other non-functional requirements that achieved a frequency of 2 and three that received a score of 1 (indicated by an asterisk). All six of these criteria were deemed worthy of inclusion in the final list. Note that *completeness* and *human factors/engineering* were not defined in IEEE Standard 24765 and will be eliminated from the final list of non-functional requirements attributes that will be considered.

A review of the definitions reveals that operability is not unique and is actually contained within the definition for availability. Based upon this, operability is removed from the list of most frequent NFRs, leaving the list with 27 unique NFRs.

[1]The on-line version of the IEEE standard was also used and is indicated by [SEVOCAB].

Table 3.15 Criteria and frequency in non-functional requirements models

Model frequency	Criteria
8	Reliability
7	Maintainability, usability,
6	Efficiency, inter-operability, portability
5	Accuracy, completeness, consistency, correctness, integrity, testability
4	Modularity, operability, reusability
3	Accessibility, adaptability, compatibility, conciseness, flexibility, human factors, modifiability, self-descriptiveness, simplicity, traceability
2	Accountability, appropriateness, augmentability, availability, clarity, communicativeness, data commonality, documentation, error tolerance, expandability, functionality, generality, independence, performance, recoverability, robustness, security, supportability, trainability, understandability
1	Aesthetics, analyzability, anomaly management, auditability, authenticity, autonomy, changeability, coexistence, cohesiveness, commonality, communications complexity, complexity, confidentiality, coupling, data structures, delivery, dependability, device efficiency, device independence, distributivity (i.e., distributedness), effectiveness, ethics, extensibility, fault-tolerance, functional overlap, functional scope, implementation, installability, learnability, legibility, legislative (i.e., legal), maturity, mean time to failure (MTTF), non-repudiation, performance efficiency, predictability, privacy, reconfigurability, replaceability, resource consumption, resource utilization, response time, safety, self-containedness, self-documentation, severity of failure, standardizability (i.e., standards), structuredness, survivability, throughput, time behavior, time performance (i.e., speed), token selection, user error protection, user interface aesthetics, verifiability, virtuality, visibility, space boundedness (i.e., space), space performance (i.e., space)

3.5.4 Notional Taxonomy of Non-functional Requirements for Systems

Twenty-seven (27) non-functional requirements have been selected for inclusion in the notional taxonomy of non-functional requirements. While this is not a complete list, it is representative of the major non-functional requirements and associated criteria that should be considered during all system design endeavors.

The 27 non-functional requirements have been arranged in the notional taxonomy or framework based upon the definitions of the criteria and how the definitions support relational concerns that are able to be represented by the collection of criteria. For example, the concern adaptation can reasonably be expected to relate to criteria such as extensibility, portability, and reusability. Similar relationships were used to construct the final framework for the NFR Taxonomy.

The final framework for the NFR Taxonomy has four concerns: (1) System Design Concerns; (2) System Adaptation Concerns; (3) System Viability Concerns;

Table 3.16 Formal definitions for most frequent NFRs

Criteria (frequency)	Formal definition
Accuracy (5)	"1. A qualitative assessment of correctness, or freedom from error. 2. A quantitative measure of the magnitude of error" IEEE and ISO/IEC (2010, p. 6)
Adaptability (3)	"Degree to which a product or system can effectively and efficiently be adapted for different or evolving hardware, software or other operational or usage environments" [SE VOCAB]
Availability (2)*	"1. The degree to which a system or component is operational and accessible when required for use. 2. Ability of a component or service to perform its required function at a stated instant or over a stated period of time" IEEE and ISO/IEC (2010, p. 29)
Compatibility (3)	"1. The ability of two or more systems or components to perform their required functions while sharing the same hardware or software environment. 2. The ability of two or more systems or components to exchange information" IEEE and ISO/IEC (2010, p. 62)
Completeness (5)	*No definition*
Conciseness (3)	"Software attributes that provide implementation of a function with a minimum amount of code" IEEE and ISO/IEC (2010, p. 69)
Consistency (5)	"1. The degree of uniformity, standardization, and freedom from contradiction among the documents or parts of a system or component. 2. Software attributes that provide uniform design and implementation techniques and notations" IEEE and ISO/IEC (2010, p. 73)
Correctness (5)	"The degree to which a system or component is free from faults in its specification, design, and implementation" IEEE and ISO/IEC (2010, p. 81)
Efficiency (6)	"1. The degree to which a system or component performs its designated functions with minimum consumption of resources. 2. Producing a result with a minimum of extraneous or redundant effort" IEEE and ISO/IEC (2010, p. 120)
Extensibility (1)*	"The ease with which a system or component can be modified to increase its storage or functional capacity. Syn: expandability, extensibility" IEEE and ISO/IEC (2010, p. 136)
Flexibility (3)	"The ease with which a system or component can be modified for use in applications or environments other than those for which it was specifically designed. Syn: adaptability cf. extendability, maintainability" IEEE and ISO/IEC (2010, p. 144)
Human factors (3)	*No definition*
Integrity (5)	"The degree to which a system or component prevents unauthorized access to, or modification of, computer programs or data" IEEE and ISO/IEC (2010, p. 181)
Interoperability (6)	"The ability of two or more systems or components to exchange information and to use the information that has been exchanged" IEEE and ISO/IEC (2010, p. 186)
Maintainability (7)	"The ease with which a hardware system or component can be retained in, or restored to, a state in which it can perform its required functions" IEEE and ISO/IEC (2010, p. 204)

(continued)

Table 3.16 (continued)

Criteria (frequency)	Formal definition
Modifiability (3)	"The ease with which a system can be changed without introducing defects cf. maintainability" IEEE and ISO/IEC (2010, p. 222)
Modularity (4)	"1. The degree to which a system or computer program is composed of discrete components such that a change to one component has minimal impact on other components. 2. Software attributes that provide a structure of highly independent components" IEEE and ISO/IEC (2010, p. 223)
Operability (4)	"The state of being able to perform the intended function" IEEE and ISO/IEC (2010, p. 240)
Portability (6)	"The ease with which a system or component can be transferred from one hardware or software environment to another" IEEE and ISO/IEC (2010, p. 261)
Reliability (8)	"The ability of a system or component to perform its required functions under stated conditions for a specified period of time" IEEE and ISO/IEC (2010, p. 297)
Reusability (4)	"The degree to which an asset can be used in more than one software system, or in building other assets" IEEE and ISO/IEC (2010, p. 307)
Robustness (2)*	"The degree to which a system or component can function correctly in the presence of invalid inputs or stressful environmental conditions cf. error tolerance, fault tolerance" IEEE and ISO/IEC (2010, p. 313)
Safety (1)*	"The expectation that a system does not, under defined conditions, lead to a state in which human life, health, property, or the environment is endangered" IEEE and ISO/IEC (2010, p. 315)
Self-descriptiveness (3)	"1. The degree to which a system or component contains enough information to explain its objectives and properties. 2. Software attributes that explain a function's implementation. cf. maintainability, testability, usability" IEEE and ISO/IEC (2010, p. 322)
Simplicity (3)	"1. The degree to which a system or component has a design and implementation that is straightforward and easy to understand. 2. Software attributes that provide implementation of functions in the most understandable manner cf. complexity" IEEE and ISO/IEC (2010, p. 327)
Survivability (1)*	"1. The degree to which a product or system continues to fulfill its mission by providing essential services in a timely manner in spite of the presence of attacks. cf. recoverability" [SE VOCAB]
Testability (4)	"1. The extent to which an objective and feasible test can be designed to determine whether a requirement is met. 2. The degree to which a requirement is stated in terms that permit establishment of test criteria and performance of tests to determine whether those criteria have been met" IEEE and ISO/IEC (2010, p. 371)
Traceability (3)	"1. The degree to which a relationship can be established between two or more products of the development process, especially products having a predecessor-successor or master-subordinate relationship to one another. 2. The identification and documentation of derivation paths (upward) and allocation or flowdown paths (downward) of work products in the work product hierarchy" IEEE and ISO/IEC (2010, p. 378)

(continued)

Table 3.16 (continued)

Criteria (frequency)	Formal definition
Understandability (2)*	"The ease with which a system can be comprehended at both the system-organizational and detailed-statement levels. NOTE Understandability has to do with the system's coherence at a more general level than readability does" IEEE and ISO/IEC (2010, p. 385)
Usability (7)	"The ease with which a user can learn to operate, prepare inputs for, and interpret outputs of a system or component" IEEE and ISO/IEC (2010, p. 388)

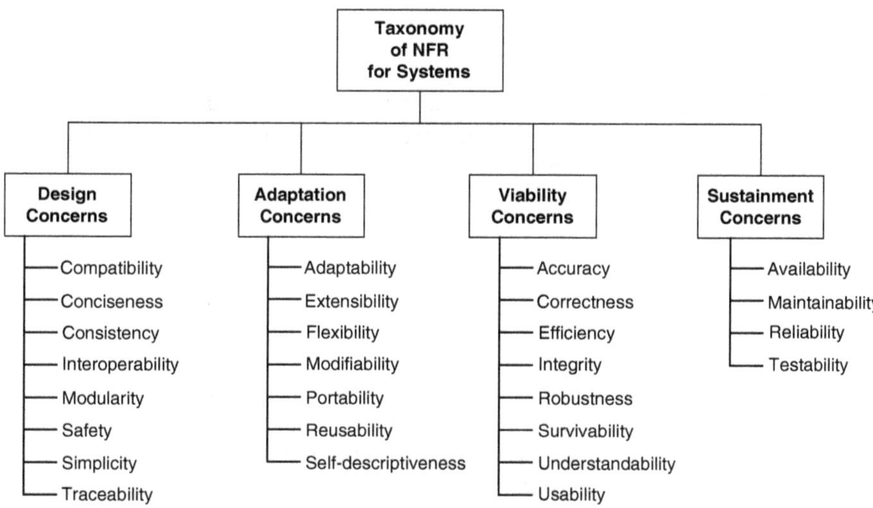

Fig. 3.5 Taxonomy of non-functional requirements for systems

and (4) System Sustainment Concerns. Figure 3.5 show the relationship between the four system concerns and the 27 non-functional requirements selected for consideration during a system's life cycle.

3.5.5 Utilization of the NFR Taxonomy for Systems

Utilization of the NFR Taxonomy for Systems requires a process for measuring the ability to achieve the non-functional requirement. By following the metrics concept articulated by Fenton and Pfleeger (1997), specific information about each attribute must be captured. A set of structural mappings that relate and individual NFR attribute from Fig. 3.5 to a specific metric and measurement entity are required. The framework for the structural mappings is based upon that described by Budgen (2003). A requisite four-level construct and example is presented in Table 3.17.

Table 3.17 Four-level structural map for Measuring NFR attributes

Level	Role	Example
Concern	A construct that permits the systems practitioner to easily group non-functional requirements based upon broad concerns	Design
Attribute	A non-functional requirement that defines one aspect of *what* a system should do	Simplicity
Metric	Measurement method or technique used to evaluate the NFR	Variety
Measurable characteristic	Specific systems characteristic that is to be measured	Number of system states

Each NFR attribute in Fig. 3.5 should have a structural map that clearly identifies the measurement method or technique and the specific systems characteristic(s) that will be used to measure the NFR.

3.6 Summary

This chapter has presented a seemingly endless array of non-functional requirements that purport to define one or more aspects of *what* a system should do. The majority of the information presented has been directed toward limiting the number of non-functional requirements a systems practitioner has to deal with during systems endeavors. While each non-functional requirement has its proponents, the discussion has to be limited in order to provide meaningful treatment of the major non-functional requirements and their associated attributes. The material in the previous sections has systematically and rationally reduced over 200 non-functional requirements to 27. As stated earlier, this may not be a complete list, but it is representative of the major non-functional requirements and associated criteria that should be considered during all system design endeavors. Furthermore, the principle of requisite parsimony (Miller 1956) has been invoked and the 27 non-functional requirements are able to be represented by a *Taxonomy of NFR for Systems* that is expressed as four concerns. Finally, a four-level framework for addressing each of the 27 non-functional requirement attributes and a measurable characteristic is presented.

Part II of the text will discuss, in two chapters, the technical details associated with non-functional requirements that are associated with systems sustainment concerns during design endeavors.

References

Blundell, J. K., Hines, M. L., & Stach, J. (1997). The measurement of software design quality. *Annals of Software Engineering, 4*(1), 235–255.

Boehm, B. W., Brown, J. R., & Lipow, M. (1976). Quantitative evaluation of software quality. In R. T. Yeh & C. V. Ramamoorthy (Eds.), *Proceedings of the 2nd International Conference on Software Engineering* (pp. 592–605). Los Alamitos, CA: IEEE Computer Society Press.

Bowen, T. P., Wigle, G. B., & Tsai, J. T. (1985). *Specification of software quality attributes: Software quality evaluation guidebook* (RADC-TR-85-37, Vol. III). Griffiss Air Force Base, NY: Rome Air Development Center.

Budgen, D. (2003). *Software design* (2nd ed.). New York: Pearson Education.

Buede, D. M. (2000). *The engineering design of systems: Models and methods*. New York: Wiley.

Cavano, J. P., & McCall, J. A. (1978). A framework for the measurement of software quality. *SIGSOFT Software Engineering Notes, 3*(5), 133–139.

Chung, L., Nixon, B. A., Yu, E. S., & Mylopoulos, J. (2000). *Non-functional requirements in software engineering*. Boston: Kluwer Academic Publishers.

Cleland-Huang, J., Settimi, R., Zou, X., & Solc, P. (2007). Automated classification of non-functional requirements. *Requirements Engineering, 12*(2), 103–120.

Cysneiros, L. M., & do Prado Leite, J. C. S. (2004). Nonfunctional requirements: From elicitation to conceptual models. *IEEE Transactions on Software Engineering, 30*(5), 328–350.

Cysneiros, L. M., & Yu, E. (2004). Non-functional requirements elicitation. In J. do Prado Leite & J. Doorn (Eds.), *Perspectives on Software Requirements* (Vol. 753, pp. 115–138). Norwell: Kluwer Academic.

de Weck, O. L., Roos, D., & Magee, C. L. (2011). *Engineering systems: Meeting human needs in a complex technological world*. Cambridge: MIT Press.

Ebert, C. (1998). Putting requirement management into praxis: dealing with nonfunctional requirements. *Information and Software Technology, 40*(3), 175–185.

Fenton, N. E., & Pfleeger, S. L. (1997). *Software metrics: A rigorous & practical approach* (2nd ed.). Boston: PWS Publications.

Grady, R. B. (1992). *Practical software metrics for project management and process improvement*. Englewood Cliffs, NJ: Prentice-Hall.

Grady, R. B., & Caswell, D. (1987). *Software metrics: Establishing a company-wide program*. Englewood Cliffs: Prentice-Hall.

Gregoriades, A., & Sutcliffe, A. (2005). Scenario-based assessment of nonfunctional requirements. *IEEE Transactions on Software Engineering, 31*(5), 392–409.

Gross, D., & Yu, E. (2001). From non-functional requirements to design through patterns. *Requirements Engineering, 6*(1), 18–36.

IEEE. (1998a). *IEEE Standard 830—IEEE recommended practice for software requirements specifications*. New York: Institute of Electrical and Electronics Engineers.

IEEE. (1998b). *IEEE Standard 1233: IEEE guide for developing system requirements specifications*. New York: Institute of Electrical and Electronics Engineers.

IEEE, & ISO/IEC. (2008). IEEE and ISO/IEC Standard 15288: Systems and software engineering—system life cycle processes. New York and Geneva: Institute of Electrical and Electronics Engineers and the International Organization for Standardization and the International Electrotechnical Commission.

IEEE, & ISO/IEC. (2010). IEEE and ISO/IEC Standard 24765: Systems and software engineering—vocabulary. New York and Geneva: Institute of Electrical and Electronics Engineers and the International Organization for Standardization and the International Electrotechnical Commission.

ISO/IEC. (1991). ISO/IEC Standard 9126: Software product evaluation—quality characteristics and guidelines for their use. Geneva: International Organization for Standardization and the International Electrotechnical Commission.

ISO/IEC. (2011). ISO/IEC Standard 25010: Systems and software engineering—Systems and software quality requirements and evaluation (SQuaRE)—system and software quality models. Geneva: International Organization for Standardization and the International Electrotechnical Commission.

Kossiakoff, A., Sweet, W. N., Seymour, S. J., & Biemer, S. M. (2011). *Systems engineering principles and practice* (2nd ed.). Hoboken: Wiley.

Mairiza, D., Zowghi, D., & Nurmuliani, N. (2010). An investigation into the notion of non-functional requirements. In *Proceedings of the 2010 ACM Symposium on Applied Computing* (pp. 311–317). New York: ACM.

McCall, J. A., & Matsumoto, M. T. (1980). *Software quality measurement manual (RADC-TR-80-109-Vol-2)*. Griffiss Air Force Base, NY: Rome Air Development Center.

Miller, G. A. (1956). The magical number seven, plus or minus two: Some limits on our capability for processing information. *Psychological Review, 63*(2), 81–97.

Nuseibeh, B. (2001). Weaving together requirements and architectures. *Computer, 34*(3), 115–119.

Pfleeger, S. L. (1998). *Software engineering: Theory and practice*. Upper Saddle River, NJ: Prentice-Hall.

Robertson, S., & Robertson, J. (2005). *Requirements-led project management*. Boston: Pearson Education.

Somerville, I. (2007). *Software engineering* (8th ed.). Boston: Pearson Education.

Suh, N. P. (2005). Complexity in engineering. *CIRP Annals—Manufacturing Technology, 54*(2), 46–63.

Sun, L., & Park, J. (2014). A process-oriented conceptual framework on non-functional requirements. In D. Zowghi & Z. Jin (Eds.), *Requirements Engineering* (Vol. 432, pp. 1–15) Berlin: Springer.

Wiegers, K. E. (2003). *Software requirements* (2nd ed.). Redmond: Microsoft Press.

Part II
Sustainment Concerns

Chapter 4
Reliability and Maintainability

Abstract Effective sustainment of systems and components during the operation and maintenance stages of the system life cycle require specific purposeful actions during the design stages of the system life cycle. The reliability and maintainability of the system and its constituent components are established as part of the system design process. Reliability and maintainability are non-functional requirements that exists at both the component- and system-level and are intertwined and interrelated. Improper reliability and maintainability designs in any level of the system's hierarchy may have far reaching affects. The ability to understand how reliability and maintainability are treated in the design process and formal metrics and measurement processes for each non-functional requirements are essential during all system design endeavors.

4.1 Introduction to Reliability and Maintainability

This chapter will address two major topics. The first topic is reliability and the second is maintainability. The first topic will review reliability and the basic theory, equations and concepts that underlie its utilization. It will then address how reliability is applied in engineering design and as a technique for determining component reliability. The section on reliability will conclude with a metric and measurable characteristic for reliability.

The second major topic of this chapter will define maintainability and discuss how it is used in engineering design. The terms used in the maintenance cycle are defined and applied to specific maintainability equations. The maintenance and support concept is introduced as an important element of the conceptual design stage of the systems life cycle. The chapter concludes with a metric and measurable characteristic for maintainability.

The chapter has a specific learning goal and associated objectives. The learning goal of this chapter is to be able to identify how the attributes of reliability and maintainability affect sustainment in systems endeavors. This chapter's goal is supported by the following objectives:

K.M. Adams, *Non-functional Requirements in Systems Analysis and Design*,
Topics in Safety, Risk, Reliability and Quality 28,
DOI 10.1007/978-3-319-18344-2_4

- Define reliability.
- Describe the reliability function and its associated probability distributions.
- Explain failure rate and the bathtub failure rate curve.
- Identify the component reliability models and their application in calculating system reliability.
- Describe the reliability processes that take place in each of the systems design phases.
- Describe how reliability is achieved in system design.
- Describe the 12 steps in a FMECA.
- Construct a structural map that relates reliability to a specific metric and measurable characteristic.
- Define maintainability.
- Identify how the maintenance and support concept is included during conceptual design.
- Describe the terminology used in the maintenance cycle.
- Construct a structural map that relates maintainability to a specific metric and measurable characteristic.
- Explain the relationship between reliability and maintainability.

The ability to achieve these objectives may be fulfilled by reviewing the materials in the chapter topics which include the following.

4.2 Reliability

Reliability is an essential characteristic of every system. Its importance is such that it is an important sub-discipline practiced within just about every major engineering discipline and is included in all engineering and system and component production endeavors. Engineers who specialize in reliability are labeled reliability engineers and Table 4.1 lists some of the scholarly journals dedicated to fundamental research in the area of reliability.

The sections that follow will not attempt to cover all aspects of reliability, but will discuss how, at a very basic level, the attributes of reliability affect sustainment in systems endeavors.

Table 4.1 Scholarly journals for reliability

Journal title	ISSN
International journal of reliability, quality, and safety engineering	0218-5393
Reliability engineering and system safety	0951-8320
Reliability—theory and applications	1932-2321
Quality and reliability engineering international	1099-1638

4.2.1 Reliability Definitions

Reliability, from a systems engineering perspective, is defined as:

> The ability of a system or component to perform its required functions under stated conditions for a specified period of time. (IEEE and ISO/IEC 2010, p. 297)

There are other definitions, presented in Table 4.2 that may contribute to improved understanding.

Dissecting these definitions, the major elements are:

- *Probability*—Fraction or a percentage specifying the number of times that one can expect an event to occur in a total number of trials.
- *Satisfactory performance*—Set of criteria to be met by a component or system.
- *Time*—A measure against which the degree of system or component performance can be related (reliability span?).
- *Specified operating conditions*—Environment in which the system or component functions.

Having a definition for and understanding the constituent elements of reliability is fine. However, what does the application of reliability engineering add to a system or component design? Simply stated, the objectives of reliability, in order of priority are:

1. To apply engineering knowledge and specialist techniques to prevent or reduce the likelihood or frequency of failures.
2. To identify and correct the causes of failures that do occur, despite the efforts to prevent them.
3. To determine ways of coping with failures that do occur, if their causes have not been corrected.

Table 4.2 Definitions for reliability

Definition	Source
"The ability or capability of the product to perform the specified functions in the designated environment for a minimum length of time or minimum number of cycles or events"	Ireson et al. (1996, p. 1.2)
"The probability that a product will operate or a service will be provided properly for a specified period of time (design life) under the design operating conditions (such as temperature, load, volt …) without failure"	Elsayed (2012, p. 3)
"The probability that an item will perform a required function without failure under stated conditions for a stated period of time"	O'Connor and Kleyner (2012, p. 1)
"The probability that the system will perform its functions correctly for a specified period of time under specified conditions"	Kossiakoff et al. (2011, p. 424)
"… The probability that a system or product will accomplish its designated mission in a satisfactory manner or, specifically, the probability that the entity will perform in a satisfactory manner for a given period when used under specified operating conditions"	Blanchard and Fabrycky (2011, p. 363)

4. To apply methods for estimating the likely reliability of new designs, and for analyzing reliability data (O'Connor and Kleyner 2012, p. 2).

Armed with a meaningful definition for reliability, and a basic set of objectives for reliability engineering, the following sections will discuss how reliability is approached during systems endeavors.

4.2.2 The Reliability Function

Reliability is related using the mathematics of probability. The basic reliability function can be related to a simple coin toss. When a coin is tossed, there are two potential outcomes, the coin lands *heads up* (*H*) or the coin lands *tails up* (*T*). The equation for this, as a function of time, *t*, is shown in Eq. 4.1.

Equation for Coin Toss

$$H(t) + T(t) = 1.0 \tag{4.1}$$

The reliability function is related similarly, with the chance of having a reliable outcome being *R*(*t*) and the chance of a failed outcome being *F*(*t*) and is shown in Eq. 4.2.

Reliability and Failure Equation

$$R(t) + \mathrm{F}(t) = 1.0 \tag{4.2}$$

Equation 4.2 may be re-written as Eq. 4.3 to show that reliability is a function of the failure rate *F*(*t*):

General Reliability Equation

$$R(t) = 1 - F(t) \tag{4.3}$$

The failure rate *F*(*t*) for a component may be represented as a probability density function (p.d.f.). By using the area under the curve for the p.d.f. By using the area under the curve for the p.d.f. Eq. 4.3 is re-written as Eq. 4.4 to show reliability using the p.d.f.

Reliability Equation with Probability Density Function

$$R(t) = 1 - \int_0^t (p.d.f)\, dt \tag{4.4}$$

A number of unique probability density functions (p.d.f.) are commonly used in reliability calculations and are presented in Table 4.3.

Table 4.3 Primary probability density function types used in reliability calculations

Probability density function type	Utilization
Poisson	The likelihood of failure is very low in a large sample size
Normal (Gaussian distribution)	The likelihood of failure is distributed equally on either side of the median failure rate
Lognormal (Galton distribution)	The likelihood of failure is expressed as the multiplicative product of many independent random variables
Weibull	The probability of failure increases over time. Use to model life data where wear out due to aging is present
Exponential	Failure rate is an exponential distribution

The expressions for each of the probability density functions may be obtained from any number of statistical handbooks. For example, an exponential p.d.f. is inserted into Eq. 4.4 the reliability equation is represented as the expression in Eq. 4.5.

Reliability Equation with Mean Life and Time Interval

$$R(t) = 1 - \int_0^t (\frac{1}{\theta} e^{-t/\theta})dt \tag{4.5}$$

where θ = mean life of the component and t = evaluation time period. To solve Eq. 4.5 two new terms are introduced: (1) failure rate (λ) which will be defined as $\lambda = 1/\theta$; and (2) mean time between failure (MTBF), defined in Table 4.4.

The solution of Eq. 4.5, using the new terms, is shown in Eq. 4.6.

Reliability Equation with MTBF and Failure Rate

$$R(t) = e^{-t/MTBF} = e^{-\lambda t} \tag{4.6}$$

When the reliability function is used to calculate the failure rate for a large population of identical components over a period of time the following behavior is observed.

> The sample experiences a high failure rate at the beginning of the operations time due to weak or substandard components, manufacturing imperfections, design errors, and installation defects. As the failed components are removed, the time between failures increases which results in a reduction in the failure rate. The period of decreasing failure rate (DFR)

Table 4.4 Definitions for MTBF and MTTF

Term	Definition
Mean time between failure (MTBF)	"The expected or observed time between consecutive failures in a system or component" IEEE and ISO/IEC (2010, p. 209)
Mean time to failure (MTTF)	"The expected time between two successive failures … when the system is nonrepairable." Elsayed (2012, p. 67)

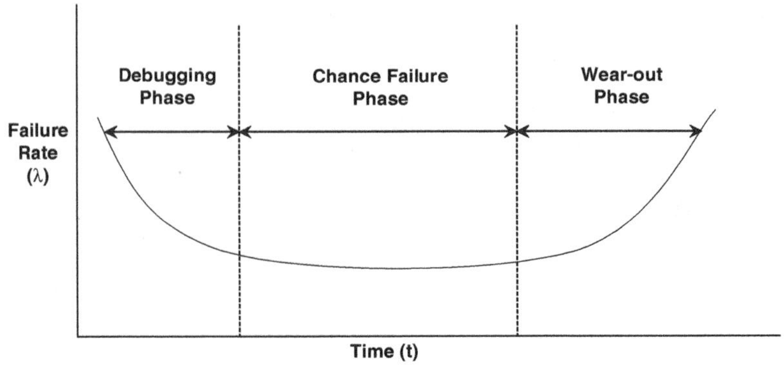

Fig. 4.1 Generic bathtub curve

> is referred to as the "infant mortality region", the "shakedown" region, the "debugging" region, or the "early failure" region. (Elsayed 2012, pp. 15–16)

The curve associated with this phenomenon is labeled the *bathtub curve* and is depicted in Fig. 4.1.

4.2.3 Component Reliability Models

As the design process moves through the conceptual and preliminary design stages the next stage is detailed design. Detailed design is where the subsystems are broken down into the required assemblies, subassemblies, components, and parts. The specific relationships and configurations of the components directly affects the reliability of the system.

Three basic relationships between components may be chosen. Components may be combined in series, in parallel, or in a combination of series and parallel relations. The sections that follow will address how reliability is calculated for each of these relationships.

4.2.3.1 Series Relationships

When components are arranged in a serial relationship, as depicted in Fig. 4.2, all components must operate satisfactorily if the system is to function as designed.

The basic reliability for the components in Fig. 4.2 is shown in Eq. 4.7.

Reliability Equation for Three Components in Series

$$R_{sys} = R_a R_b R_c \tag{4.7}$$

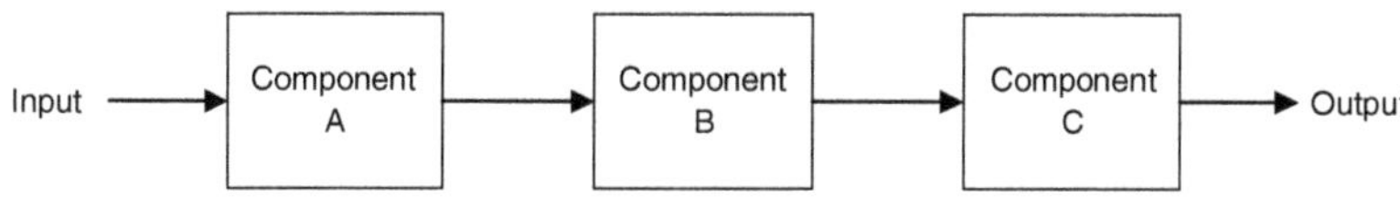

Fig. 4.2 Serial relationship between system components

By inserting Eq. 4.6 into Eq. 4.7 the failure rate is introduced into the reliability equation as depicted in Eq. 4.8.

Reliability Equation Using Failure Rate for Three Components in Series

$$R_{sys} = (e^{-\lambda_a t})(e^{-\lambda_b t})(e^{-\lambda_c t}) \tag{4.8}$$

As an example, imagine that Component A is a radio frequency receiver, Component B is an amplifier, and Component C is a radio frequency transmitter. The receiver reliability is 0.9512 and has an MTBF of 6000 h, the amplifier reliability is 0.9821 and has an MTBF of 4500 h, and the transmitter reliability is 0.9357 with an MTBF of 10,000 h. The system is expected to operate for 1000 h.

The overall reliability of the system is calculated using Eq. 4.8.

$$R_{sys} = (0.9512)(0.9821)(0.9357) = 0.8741$$

The failure rate for a component is the inverse of the MTBF, as shown in Eq. 4.9.

Failure Rate and MTBF

$$\lambda_x = \frac{1}{MTBF_x} \tag{4.9}$$

The failure rate for each of the components is as follows:

$$\lambda_A = \frac{1}{6000} = 0.000167 \text{ failures per hour}$$

$$\lambda_B = \frac{1}{4500} = 0.000222 \text{ failures per hour}$$

$$\lambda_B = \frac{1}{10000} = 0.000100 \text{ failures per hour}$$

By using Eq. 4.8 the overall reliability of the system over a period of 1000 h is:

$$R_{sys} = (e^{-\lambda_a t})(e^{-\lambda_b t})(e^{-\lambda_c t})$$

$$R_{sys} = \left(e^{-(0.000167)(1000)}\right)\left(e^{-(0.000222)(1000)}\right)\left(e^{-(0.000100)(1000)}\right)$$

$$R_{sys} = \left(e^{-(0.167)}\right)\left(e^{-(0.222)}\right)\left(e^{-(0.01)}\right)$$

$$R_{sys} = \left(e^{-(0.399)}\right)$$

$$R_{sys} = 0.601$$

This shows that the system, as configured, has a 60.1 % probability of surviving to 1000 h.

4.2.3.2 Parallel Relationships

When components are arranged in a parallel relationship, as depicted in Fig. 4.3, all components must fail to cause a total system failure.

The basic reliability for the components A and B in Fig. 4.3 is shown in Eq. 4.10.

Reliability Equation for Two Components in Parallel

$$R_{sys} = R_a + R_b - (R_a)(R_b) \tag{4.10}$$

And the reliability for components A, B and C is shown in Eq. 4.11.

Reliability Equation for Three Components in Parallel

$$R_{sys} = 1 - [(1 - R_a)(1 - R_b)(1 - R_c)] \tag{4.11}$$

As an example, imagine a system constructed where components A, B, and C are identical power supplies with a reliability of 0.975. If only two components are in

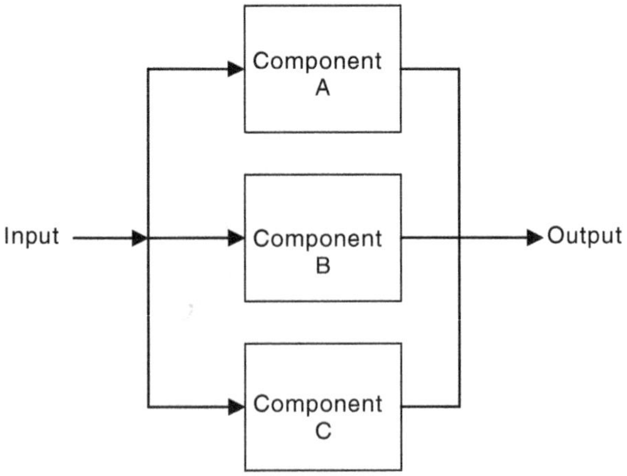

Fig. 4.3 Parallel relationship between system components

the design, A and B, which is a system with two identical power supplies, the corresponding system reliability is calculated as:

$$R_{sys} = 0.975 + 0.975 - [(0.975)(0.975)] = 0.99375$$

If a third identical power supply is added in parallel to this system design, then the reliability of the system increases:

$$R_{sys} = 1 - [(1 - 0.975)(1 - 0.975)(1 - 0.975)] = 0.999984$$

4.2.3.3 Combined Series-Parallel Relationships

When components are arranged in relationships that combine both series and parallel relations, as depicted in Fig. 4.4, a variety of failure combinations may cause the system to fail.

The basic reliability for the components A, B, C and D in Fig. 4.4 is shown in Eq. 4.12.

Reliability Equation for Components in a Combined Series-Parallel Relationship

$$R_{sys} = [1 - [(1 - R_a)(1 - R_b)][1 - [(1 - R_c)(1 - R_d)] \tag{4.12}$$

The number of possible combinations of components in a system is endless. However, the system's configuration for reliability is a purposeful element of the design process aimed at achieving required reliability requirements. More detailed treatment of reliability may be found in focused texts on Reliability Engineering by Elsayed (2012) and O'Connor and Kleyner (2012).

The next section will discuss how reliability is addressed during systems design efforts.

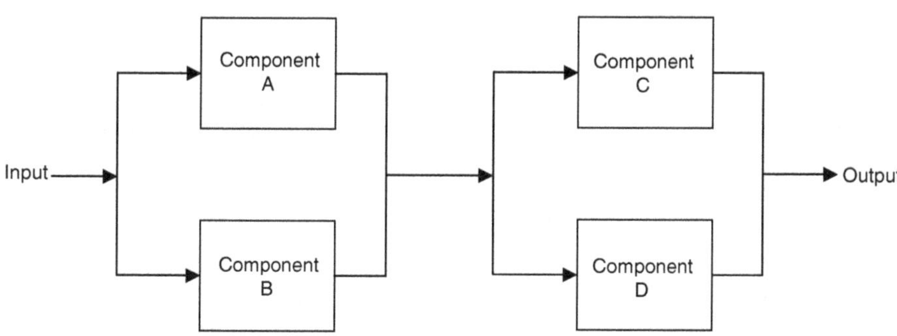

Fig. 4.4 A combined series-parallel relationship between system components

4.2.4 Reliability in System Design Efforts

Reliability is a major factor in determining system effectiveness. Reliability is most often defined within: (1) operational requirements (i.e., availability) and maintenance concepts (i.e., maintainability); (2) system requirements; and (3) performance factors.

IEEE Standard 1220—Systems engineering—Application and management of the systems engineering process (IEEE 2005) specifically addresses reliability in three areas.

- As an element of the requirements analysis task in Section 6.1.5
  During the requirements analysis task, system effectiveness measures are identified that reflect the overall stakeholder expectations. Reliability is a key Measure of Effectiveness (MOE) that should be identified in this task.
- As an element of the design synthesis task in Section 6.5.2
  Design solution alternatives are evaluated and the non-functional requirement attribute reliability is a distinct measure used in discrimination of the alternatives.
- As an element of the design synthesis task in Section 6.5.12.

> … assesses failure modes, the effects, and the criticality of failure for design alternatives. The hardware, software, and human elements of the design alternatives should be analyzed, and historical or test data should be applied, to refine an estimate of the probability of successful performance of each alternative. A failure modes and effects analysis (FMEA) should be used to identify the strengths and weaknesses of the design solution. For critical failures, the project conducts a criticality analysis to prioritize each alternative by its criticality rating. The results of this analysis are used to direct further design efforts to accommodate redundancy and to support graceful system degradation. (IEEE 2005, p. 52)

The section that follows will discuss the Failure Mode and Effect Analysis (FMEA) mentioned in IEEE Standard 1220.

4.2.5 FMEA and FEMCA

Failure Mode and Effect Analysis (FMEA) is defined as:

> An analytical procedure in which each potential failure mode in every component of a product is analyzed to determine its effect on the reliability of that component and, by itself or in combination with other possible failure modes, on the reliability of the product or system and on the required function of the component; or the examination of a product (at the system and/or lower levels) for all ways that a failure may occur. For each potential failure, an estimate is made of its effect on the total system and of its impact. In addition, a review is undertaken of the action planned to minimize the probability of failure and to minimize its effects. (IEEE and ISO/IEC 2010, p. 139)

FMEA is also known as FMECA or Failure Modes, Effects, and Criticality Analysis and is a popular and widely used reliability design technique. FMEA may be applied to functional or physical entities and may start as early as the conceptual design stage in the systems life cycle. The FMEA/FMECA goals are to:

- Design process to improve inherent system reliability.
- Identifies potential system weaknesses.

The process uses a *bottom-up* approach where the analysis considers failures at the lowest level of the system's hierarchy and moves upward to determine effects at the higher levels in the hierarchy. The process traditionally uses 12 steps:

1. Define the system, outcomes, and technical performance measures
2. Define the system in functional terms
3. Do a top-down breakout of system level requirements
4. Identify failure modes, element by element
5. Determine causes of failure
6. Determine effects of failure
7. Identify failure/defect detection means
8. Rate failure mode severity
9. Rate failure mode frequency
10. Rate failure mode detection probability (based on item #7)
11. Analyze failure mode criticality where criticality is a function of severity (#8), frequency (#9), and probability of detection (#10) as expressed in a Risk Priority Number (RPN)
12. Initiate recommendations for improvement.

Bernstein (1985) provides an example of a FMECA for a mechanical system that should be reviewed for understanding. Finally, there is an established international standard for conducting a FMEA/FMECA which is contained in *IEC Standard 60812 Analysis Techniques for System Reliability—Procedure for Failure Mode and Effects Analysis (FMEA)* (IEC 2006).

The next section will discuss how to measure reliability.

4.2.6 Measuring Reliability

At the end of Chap. 3 the importance of being able to measure each non-functional attribute was stressed. A structural mapping that relates reliability to a specific metric and measurement entity are required. The four-level construct for reliability is presented in Table 4.5.

The section that follows will discuss the non-functional attribute for maintainability.

Table 4.5 Four-level structural map for measuring reliability

Level	Role
Concern	Systems sustainment
Attribute	Reliability
Metric	Component reliability
Measurable characteristic	Mean time between failure (MTBF) or mean time to failure (MTTF)

4.3 Maintainability

This section will review the basics of maintainability and how it is applied during systems endeavors. Maintainability is closely related to reliability and is a central element of sustainment.

4.3.1 Maintainability Definitions

Maintainability, from a systems engineering perspective, is defined as:

> The ease with which a hardware system or component can be retained in, or restored to, a state in which it can perform its required functions. (IEEE and ISO/IEC 2010, p. 204)

There are other definitions, presented in Table 4.6 that may contribute to improved understanding.

Dissecting these definitions, the major elements are:

- *Maintain*—The ability to take the actions necessary to keep the system in a fully operable condition.
- *Maintenance*—"the process of retaining a hardware system or component in, or restoring it to, a state in which it can perform its required functions" (IEEE and ISO/IEC 2010, p. 205).

Table 4.6 Definitions for maintainability

Definition	Source
"A measure of the ease of accomplishing the functions required to maintain the system in a fully operable condition"	Kossiakoff et al. (2011, p. 428)
"That characteristic of design and installation that reflects the ease, accuracy, safety, and economy of performing maintenance actions"	Blanchard and Fabrycky (2011, p. 112)
"The ability of a system to be maintained"	Blanchard and Fabrycky (2011, p. 411)

Having a definition for and understanding the constituent elements of maintainability is an essential step in understanding how maintainability can be measured.

The next section will discuss the element of maintainability.

4.3.2 Terminology Associated with Maintainability

The first element of maintainability to understand is maintenance. Maintenance is generally classified as either corrective or preventive.

- *Corrective*: unscheduled, caused by failure, required to restore system level of performance.
- *Preventive*: scheduled, designed to prevent failure, required to retain system level of performance.

Both of the maintenance types involve a number of specific actions in order to affect the actual maintenance of the system and its components. The basic elements of a corrective maintenance cycle for a component or system are depicted in Fig. 4.5.

The terms associated with the corrective maintenance cycle are defined in Table 4.7.

From the relationships depicted in Fig. 4.5 the maintenance terms are related by Eq. 4.13.

Basic Maintenance Downtime Equation

$$MDT = M + LDT + ADT \tag{4.13}$$

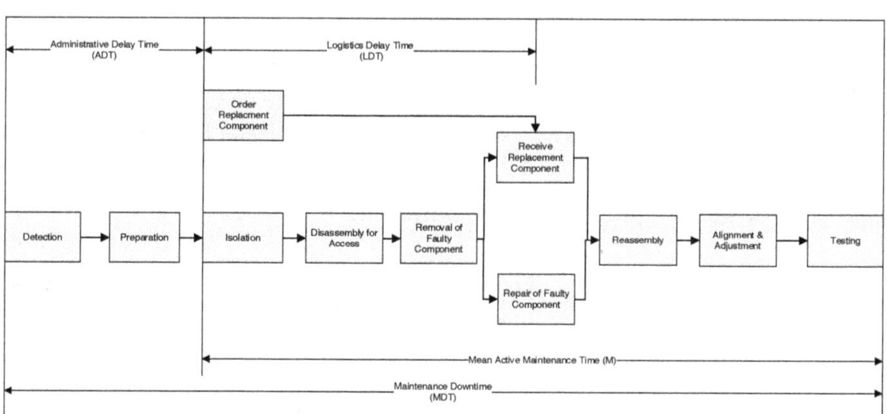

Fig. 4.5 Corrective maintenance cycle

Table 4.7 Terms in corrective maintenance cycle

Term	Definition
Administrative delay time (ADT)	That portion of the downtime during which maintenance is delayed due to planning, personnel assignment, labor issues, constraints, etc
Logistics delay time (LDT)	That portion of the downtime waiting for material, a piece of special test equipment, tool, facility, or transportation
Maintenance downtime (MDT)	The total period of time required, during which the system is non-operational, to repair and restore a system to full operating status
Mean active maintenance time (M)	The total period of time required, during which the system is non-operational, to repair and restore a system to full operating status. Is expressed as a function of both Mct, Mpt, and fpt (frequency of preventive maintenance action)
Mean corrective maintenance time (Mct)	Average elapsed time for unscheduled maintenance
Mean preventive maintenance time (Mpt)	Average elapsed time for preventive or scheduled maintenance

Knowing that the Mean Active Maintenance Time (M) is the sum of the mean corrective and mean preventive maintenance times, Eq. 4.13 may be expressed as Eq. 4.14:

Expanded Maintenance Downtime Equation

$$MDT = Mct + Mpt + LDT + ADT \tag{4.14}$$

4.3.3 *Maintainability Calculations*

Each component has a measure of maintainability designated as the Mean Time Between Maintenance (MTBM), which is the mean or average time between all maintenance actions knowing that there are both unscheduled corrective ($MTBM_u$) and scheduled preventive ($MTBM_s$) maintenance actions. The formula for MTBM is shown in Eq. 4.15.

Equation for Mean Time between Maintenance

$$MTBM = \frac{1}{\frac{1}{MTBM_u} + \frac{1}{MTBM_s}} \tag{4.15}$$

Both of these measures are primary measures for maintainability and will be required to address the non-functional requirement attribute for availability in the next chapter.

4.3.4 Maintenance Support Concept

Because maintenance is an important element of during system operation, it must be addressed early in the system design process. During conceptual design the design team will be tasked with considering specific concepts that support the full range of activities that will occur during the systems life cycle. IEEE Standard 1220 requires the definition of life cycle process concepts, which includes the maintenance support concept:

> Develop, produce, test, distribute, operate, support, train, and dispose of system products under development. (IEEE 2005, p. 40)

Much of the design team's focus should be on the cost drivers and higher risk elements that are anticipated to impact the system throughout its life cycle. Because system operations and maintenance activities consume large portions of a systems total life cycle costs, the maintenance support concept is particularly important. A maintenance concept may be defined as:

> The set of various maintenance interventions (corrective, preventive, condition based, etc.) and the general structure in which these interventions are foreseen. (Waeyenbergh and Pintelon 2002, p. 299)

Figure 4.6 depicts some of the major elements in a systems maintenance and support concept.

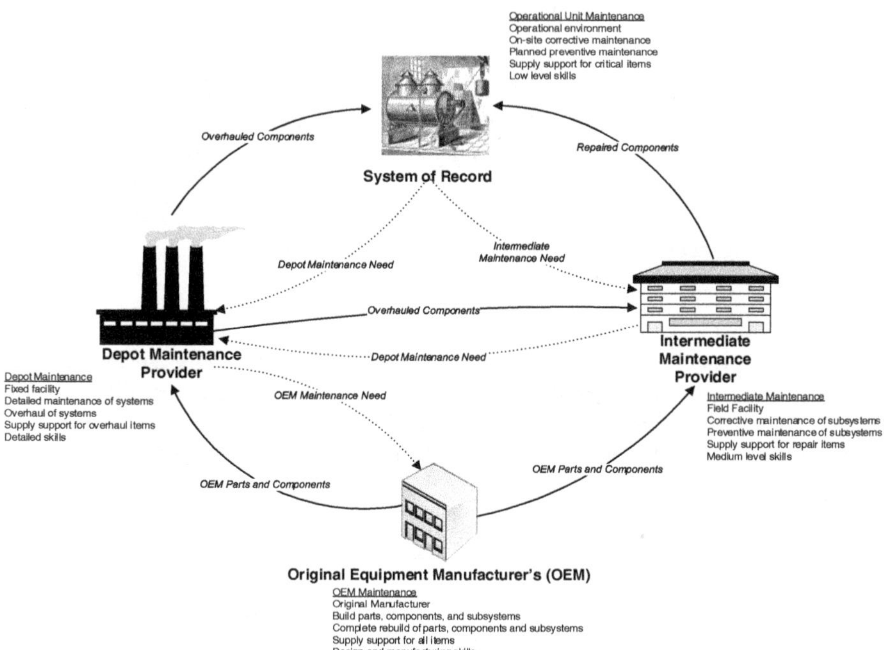

Fig. 4.6 Maintenance and support concept

Table 4.8 Maintenance and support considerations

Support item	Support item considerations
1. Levels of maintenance	How many levels are included in the maintenance concept is a function of the anticipated frequency of the maintenance, the task complexity, skill level requirements, and facility needs
2. Repair policies	Part and component failure policies should specify whether the item is non-repairable, partially repairable, or fully repairable and the organizational level-of-repair
3. Organizational responsibilities	Maintenance may be assigned to the operational facility or to any of the supporting organizations shown on the previous page
4. Maintenance support elements	Inventory levels for maintenance parts, repair parts, overhaul parts and provisioning. Special support equipment (facilities, equipment, tools). Training requirements
5. Effectiveness requirements	Predicted part and component failure rates will drive the spare-part demand rate
6. Environment	The environment that surrounds the system. This includes understanding vibration, shock, temperature, humidity, noise, pressures, cycles, etc

Areas that should be addressed are presented in Table 4.8.

The next section will address how maintainability is addressed during systems design efforts.

4.3.5 Maintainability in Systems Design Efforts

Maintainability is a major factor in determining system effectiveness. Maintainability is most often defined within: (1) operational requirements (i.e., availability) and maintenance concepts (i.e., maintainability); (2) system requirements; and (3) performance factors.

IEEE Standard 1220—Systems engineering—Application and management of the systems engineering process (IEEE 2005) specifically addresses reliability in three areas.

- As an element of the requirements analysis task in Section 6.1.5
  During the requirements analysis task system effectiveness measures are identified that reflect the overall stakeholder expectations. Maintainability is a key Measure of Effectiveness (MOE) identified in this task.
- As an element of the requirements analysis task in Section 6.1.9
  The design team must define life cycle process concepts, which includes the maintenance support concept.
- As an element of the design synthesis task in Section 6.5.2
  Design solution alternatives are evaluated and the non-functional requirement attribute maintainability is a distinct measure used in the discriminating of the alternatives.

The next section will discuss how to measure maintainability.

Table 4.9 Four-level structural map for measuring maintainability

Level	Role
Concern	Systems sustainment
Attribute	Maintainability
Metric	Component maintainability
Measurable characteristic	Mean time between unscheduled corrective maintenance ($MTBM_u$) or mean time between scheduled preventive maintenance ($MTBM_s$)

4.3.6 Measuring Maintainability

At the end of Chap. 3 the importance of being able to measure each non-functional attribute was stressed. To support this, a structural mapping that relates maintainability to a specific metric and measurement entity are required. The four-level construct for maintainability is presented in Table 4.9.

4.4 Summary

In this chapter the non-functional requirements for reliability and maintainability have been reviewed. In each case a formal definition has been provided along with additional explanatory definitions, terms, and equations. The ability to purposefully account for the non-functional requirement during the design process has also been addressed. Finally, a formal metric and measurement characteristic have been proposed for evaluating each non-functional requirement attribute.

The chapter that follows will discuss additional sustainment concerns and address the non-functional attributes for availability and operability, and testability.

References

Bernstein, N. (1985). Reliability analysis techniques for mechanical systems. *Quality and Reliability Engineering International, 1*(4), 235–248.

Blanchard, B. S., & Fabrycky, W. J. (2011). *Systems engineering and analysis* (5th ed.). Upper Saddle River: Prentice-Hall.

Elsayed, E. A. (2012). *Reliability engineering* (2nd ed.). Hoboken: Wiley.

IEC. (2006). *IEC Standard 60812: Analysis techniques for system reliability—procedure for failure mode and effects analysis (FMEA)*. Geneva: International Electrotechnical Commission.

IEEE. (2005). *IEEE Standard 1220: Systems engineering—application and management of the systems engineering process*. New York: Institute of Electrical and Electronics Engineers.

IEEE, & ISO/IEC. (2010). *IEEE and ISO/IEC Standard 24765: Systems and Software Engineering —Vocabulary*. New York and Geneva: Institute of Electrical and Electronics Engineers and the International Organization for Standardization and the International Electrotechnical Commission.

Ireson, W. G., Coombs, C. F., & Moss, R. Y. (1996). *Handbook of reliability engineering and management* (2nd ed.). New York: McGraw-Hill.

Kossiakoff, A., Sweet, W. N., Seymour, S. J., & Biemer, S. M. (2011). *Systems engineering principles and practice* (2nd ed.). Hoboken: Wiley.

O'Connor, P. D. T., & Kleyner, A. (2012). *Practical reliability engineering* (5th ed.). West Sussex: Wiley.

Waeyenbergh, G., & Pintelon, L. (2002). A framework for maintenance concept development. *International Journal of Production Economics, 77*(3), 299–313.

Chapter 5
Availability, Operability, and Testability

Abstract Effective sustainment of systems and components during the operation and maintenance stages of the system life cycle require specific purposeful actions during the design stages of the system life cycle. The availability and testability of the system and its constituent components are essential to ensure that the systems continues to provide the required functions to its stakeholders. Availability and testability are non-functional requirements that exists at both the component- and system-level and are intertwined and interrelated. Inattention to testability concerns may lead to decreased availability having far reaching affects that include the systems own viability. The ability to understand how availability and testability are implemented in the design process and formal metrics and measurement processes for each non-functional requirements ensure that they are adequately addressed during all system design endeavors.

5.1 Introduction to Availability and Testability

This chapter will address two major topics. The first topic is availability and the second is testability. The first topic will review availability and the basic theory, equations and concepts that underlie its utilization. A section will address how availability is applied in engineering design and conclude with a metric and measureable characteristic for reliability.

The second major topic of this chapter will define testability and discuss how it is used in engineering design. The relationship to availability is reviewed and applied to the availability equation. The section concludes with a metric and measureable characteristic for testability.

The chapter has a specific learning goal and associated objectives. The learning goal of this chapter is to be able to identify how the attributes of availability and testability affect sustainment in systems endeavors. This chapter's goal is supported by the following objectives:

K.M. Adams, *Non-functional Requirements in Systems Analysis and Design*,
Topics in Safety, Risk, Reliability and Quality 28,
DOI 10.1007/978-3-319-18344-2_5

- Define availability.
- Describe the terminology used to calculate availability.
- Describe the relationship between maintainability and availability.
- Construct a structural map that relates availability to a specific metric and measurable characteristic.
- Define testability.
- Describe the relationship between testability and operational availability.
- Construct a structural map that relates testability to a specific metric and measurable characteristic.
- Explain the relationship between testability and availability.

The ability to achieve these objectives may be fulfilled by reviewing the materials in the sections that follow.

5.2 Availability and Operability

This section will review the basics of availability and how it is applied during systems endeavors. Availability is an important measure utilized in assessing a systems' ability to provide the required functions and services to its stakeholders.

5.2.1 Availability and Operability Definitions

Operability, from a systems engineering perspective, is defined as

> The state of being able to perform the intended function (IEEE and ISO/IEC 2010, p. 240).

Availability, from a systems engineering perspective, is defined as:

> 1. The degree to which a system or component is operational and accessible when required for use. 2. Ability of a component or service to perform its required function at a stated instant or over a stated period of time (IEEE and ISO/IEC 2010, p. 29).

From these definitions it should be clear that the non-functional requirement for operability is easily satisfied within the definition for availability. As a result, the rest of this chapter will treat operability as part of the non-functional requirement for availability.

Availability is usually simply stated as the ratio of the system uptime over the sum of system uptime and system downtime. Availability as a general concept is "the period of time for which an asset is capable of performing its specified function, expressed as a percentage" (Campbell 1995, p. 174).

However, availability has a number of unique definitions, characterized by either (1) the time interval being considered, or (2) the type of downtime (i.e., either corrective repairs or scheduled maintenance). There are other definitions presented in Table 5.1 that may contribute to improved understanding.

Table 5.1 Key elements that differentiate the definitions for availability

Definition	Source
Inherent Availability (**A_i**): "Includes only the corrective maintenance of the system (the time to repair or replace the failed component), and excludes ready time, preventive maintenance downtime, logistics (supply) time, and waiting administrative time"	Elsayed (2012, p. 202)
Achieved Availability (**A_a**): "Includes corrective and preventive maintenance downtime. It is expressed as a function of the frequency of maintenance, and the mean maintenance time"	Elsayed (2012, p. 202)
Operational Availability (**A_o**): "The repair time includes many elements: the direct time of maintenance and repair and the indirect time which includes ready time, logistics time, and waiting or administrative downtime"	Elsayed (2012, p. 203)

Operational availability (A_o) is the most appropriate measure of system or component availability since it includes systems and their components in their real-world operational environments. The next section will address operational availability (A_o) mathematically.

5.2.2 Equations for Operational Availability (A_o)

Operational availability may be simply expressed as shown in Eq. 5.1.

Basic Availability Equation

$$A_o = \frac{System\ uptime}{System\ total\ time\ (uptime + downtime)} \tag{5.1}$$

Equation 5.1 may be expanded by including the maintainability terms *mean time between failure* (MTBF) and *mean time to repair* (MTTR) that were discussed in Chap. 4 and by introducing a new term *mean logistics delay time* (MLDT). The equation for operational availability is expanded and shown in Eq. 5.2.

Expanded Availability Equation

$$A_O = \frac{MTBF}{MTBF + MTTR + MLDT} \tag{5.2}$$

MLDT includes both *mean supply delay time* (MSDT), *mean outside assistance delay time* (MOADT) and *mean administrative delay time* (MADT). The terms are defined as follows:

- MSDT includes delays during the acquisition of spare parts, test equipment, and special tooling required to accomplish the maintenance.
- MOADT includes delays due to the arrival of specialized maintenance personnel during travel to the systems' operational location to perform maintenance.

- MADT includes delays due to the development of procedures, permission to restrict system operation during maintenance, system isolation, and preparation for and the setting conditions (i.e., draining fluids, de-energizing circuits, etc.) within the system that permit maintenance actions to be accomplished.

Using these terms the MLDT is expanded and shown in Eq. 5.3.

Mean Logistics Delay Time Equation

$$MLDT = MSDT + MOADT + MADT \tag{5.3}$$

The operational availability (A_o) equation may now be re-written to include the additional terms from MLDT and is shown in Eq. 5.4.

Fully Expanded Availability Equation

$$A_0 = \frac{MTBF}{MTBF + MTTR + MSDT + MOADT + MADT} \tag{5.4}$$

5.2.3 Availability in Systems Design Efforts

Availability, like reliability and maintainability, is a major factor in determining system effectiveness. Availability is addressed in *IEEE Standard 1220—Systems engineering—Application and management of the systems engineering process* (IEEE 2005) and is specifically addressed in three areas.

- As an element of the requirements analysis task in Sect. 6.1.1
  Stakeholder expectations are defined and balanced in terms that address specific non-functional requirements such as availability.
- As an element of the requirements analysis task in Sect. 6.1.9
  The design team must define life cycle process concepts, which includes the maintenance support concept which will have a direct effect on availability.
- As an element of the design synthesis task in Sect. 6.5.2
  Design solution alternatives are evaluated and the non-functional requirement attribute availability is a distinct measure used in the discriminating of the alternatives.

A typical stakeholder requirement may state that *the system must be available at least 99.5 percent of the time*. By stating the required uptime the stakeholders are constraining the amount of downtime and the variables that contribute to downtime (i.e., MTBF, MTTR, and MLDT (i.e., MSDT, MOADT, and MADT)).

The next section will discuss how to measure availability.

Table 5.2 Four-level structural map for measuring availability

Level	Role
Concern	Systems Sustainment
Attribute	Availability
Metric	Component availability
Measurable characteristic	Mean time between failure (MTBF), Mean time to repair (MTTR), and mean logistic delay time (MLDT)

5.2.4 Measuring Operational Availability (A_o)

At the end of Chap. 3 we stressed the importance of being able to measure each non-functional attribute. A structural mapping that relates availability to a specific metric and measurement entity are required. The four-level construct for availability is presented in Table 5.2.

The section that follows will discuss an additional sustainment concern by addressing the non-functional attribute for testability.

5.3 Testability

In this section the basics of testability and how it is applied during systems endeavors will be reviewed. Testability is an emerging measure that could be utilized as a means for improving the ability to properly assess a system's conformance with the functions and services required by its stakeholders.

5.3.1 Testability Definitions

Testability, from a systems engineering perspective, is defined as

> 1. The extent to which an objective and feasible test can be designed to determine whether a requirement is met. 2. The degree to which a requirement is stated in terms that permit establishment of test criteria and performance of tests to determine whether those criteria have been met (IEEE and ISO/IEC 2010, p. 371).

Testability, as a non-functional requirement, has additional definitions shown in Table 5.3 that may provide improved understanding of the term and its use in systems design endeavors.

Using the oldest definition for testability, Valstar (1965) posits that testability is an element of the inherent availability (A_i) and can be related by the formula in Eq. 5.5.

Table 5.3 Additional definitions for testability

Definition	Source
"Testability, symbolized by τ, the mean-time-to-test or to find a failure provided that test equipment, facilities and manpower are available"	Valstar (1965, p. 54)
"The capability of a system that allows effective fault detection and diagnosis"	Kelley et al. (1990, p. 22)
"The tendency for failures to be observed during testing when faults are present. Software has high testability if it tends to expose faults during testing, producing failures for many inputs that execute a fault. Software had low testability if it tends to hide faults during testing, producing almost no failures even through at least one fault is present"	Voas and Miller (1993, p. 207)
"Testability can suggest places where faults can hide from testing, which testing cannot do. Testability complements formal verification by providing empirical evidence of behavior, which formal verification cannot do"	Voas and Miller (1995, p. 18)

Inherent Availability Equation

$$A_i = \frac{\omega}{\omega + \tau + \rho} \tag{5.5}$$

where

ω Mean time to failure or MTTF.

τ Mean time to find a failure provided that test equipment, facilities, and manpower are available.

ρ Repairability or mean time to repair or MTTR.

Re-written using the more familiar terms from the previous section, Eq. 5.5 becomes Eq. 5.6.

Inherent Availability Equation with MTTF and MTTR

$$A_i = \frac{MTTF}{MTTF + \tau + MTTR} \tag{5.6}$$

Because inherent availability considers only corrective maintenance, that is, faults that are caused by failure, testability (τ) also has a role in the prediction of the operational availability of a system. The operational availability Eq. 5.5 may be re-written as shown in Eq. 5.7.

Expanded Availability Equation with Testability

$$A_0 = \frac{MTBF}{MTBF + \tau + MTTR + MSDT + MOADT + MADT} \tag{5.7}$$

In summary, testability directly affects system availability.

> Availability is dependent on how well the operator can assess the condition of the system, how quickly he/she can detect and locate the cause of degraded or failed units, and how efficiently he/she can rectify the malfunction (Kelley et al. 1990, p. 22).

5.3.2 Testability in Systems Design

Testability may be categorized in two ways:

1. *Architectural testability*—"an evaluation of the characteristics of a system based on its design features and the specified intent of the design" (Kelley et al. 1990, p. 22).
2. *Modal testability*—"an evaluation of the testability of a system's design while configured to perform one or more specific functions. This type of analysis is based on the system's design features, its functional flow characteristics, and its performance specifications" (Kelley et al. 1990, p. 22).

Formal measures for testability attributes have been proposed for a number of design tasks and are presented in Table 5.4.

Design for testability is a technique whereby designers proactively ensure that their design decisions support the development of a robust test program throughout the systems life cycle (Buschmann 2011). For both cases in Table 5.4 the tasks and measures describe correlated testing criterion that brings a basis for evaluating design as part of a broader design for testability approach. By having a defined testability metric and associated characteristic that are measureable, design alternatives may be afforded another measure with which they may be discriminated.

Testability, like availability, is a major factor in determining system effectiveness. Availability is addressed in *IEEE Standard 1220—Systems engineering—Application and management of the systems engineering process* (IEEE 2005) specifically addresses testability in three areas.

- As an element of the synthesis task in Sects. 6.5.4 and 6.5.13
  Determine the degree to which testability has been included in the solutions.
  Assess the testability of design alternatives to determine built-in test (BIT) and/or fault isolation test (FIT) requirements.

Table 5.4 Examples of testability measures in design tasks

Design task	Measure	Source
Architecture development	Measures of the relative controllability and observability of signals in the system architecture. The testability information is derived from reachability graph analysis of the corresponding Petri Net representation of the system architecture	Jiang et al. (2000)
UML Class Diagram	Measure the number and complexity of the interactions due to polymorphic uses	Baudry et al. (2002)

Table 5.5 Four-level structural map for measuring testability

Level	Role
Concern	Systems Sustainment
Attribute	Testability
Metric	Component testability
Measurable characteristic	τ, the Mean time to find a failure provided that test equipment, facilities, and manpower are available

5.3.3 Measuring Testability

At the end of Chap. 3 the importance of being able to measure each non-functional attribute was stressed. In support of this importance, a structural mapping that relates testability to a specific metric and measurement entity are required. The four-level construct for testability is presented in Table 5.5.

5.4 Summary

In this chapter the non-functional requirements for availability and testability have been reviewed. In each case a formal definition has been provided along with additional explanatory definitions, terms, and equations. The ability to effect the non-functional requirement during the design process has also been addressed. Finally, a formal metric and measurement characteristic have been proposed for evaluating each non-functional requirement attribute.

The next Part of the text will shift the focus to concerns that are directly related to the design itself. The first chapter in the Part on Design Concerns will address the non-functional attributes for conciseness, modularity, simplicity, and traceability. The second chapter in Part III on Design Concerns will address the non-functional attributes for compatibility, consistency, interoperability, and safety.

References

Baudry, B., Le Traon, Y., & Sunye, G. (2002). Testability analysis of a UML class diagram. *Proceedings of the Eighth IEEE Symposium on Software Metrics* (pp. 54–63). Los Alamitos, CA: IEEE Computer Society.

Buschmann, F. (2011). Tests: The Architect's best friend. *IEEE Software, 28*(3), 7–9.

Campbell, J. D. (1995). *Uptime: Strategies for excellence in maintenance management*. Portland, OR: Productivity Press.

Elsayed, E. A. (2012). *Reliability engineering* (2nd ed.). Hoboken, NJ: Wiley.

IEEE. (2005). *IEEE Standard 1220: Systems engineering—Application and management of the systems engineering process*. New York: Institute of Electrical and Electronics Engineers.

IEEE, and ISO/IEC. (2010). *IEEE and ISO/IEC Standard 24765: Systems and software engineering—Vocabulary*. New York and Geneva: Institute of Electrical and Electronics Engineers and the International Organization for Standardization and the International Electrotechnical Commission.

Jiang, T., Klenke, R. H., Aylor, J. H., & Gang, H. (2000). System level testability analysis using Petri nets. *Proceedings of the IEEE International High-Level Design Validation and Test Workshop* (pp. 112–117). Los Alamitos, CA: IEEE Computer Society.

Kelley, B. A., D'Urso, E., Reyes, R., & Treffner, T. (1990). System testability analyses in the Space Station Freedom program. *Proceedings of the IEEE/AIAA/NASA 9th Digital Avionics Systems Conference* (pp. 21–26). Los Alamitos, CA: IEEE Computer Society.

Valstar, J. E. (1965). The contribution of testability to the cost-effectiveness of a weapon system. *IEEE Transactions on Aerospace, AS-3*(1), 52–59.

Voas, J. M., & Miller, K. W. (1993). Semantic metrics for software testability. *Journal of Systems and Software, 20*(3), 207–216.

Voas, J. M., & Miller, K. W. (1995). Software testability: The new verification. *IEEE Software, 12*(3), 17–28.

Part III
Design Concerns

Chapter 6
Conciseness, Modularity, Simplicity and Traceability

Abstract The design of systems and components during the design stage of the systems life cycle requires specific purposeful actions to ensure effective designs and viable systems. Designers are faced with a number of *design concerns* that they must embed into the design in every instance of thinking and documentation. Four of these concerns are addressed by the non-functional requirements for conciseness, modularity, simplicity, and traceability. Formal understanding of each of these non-functional requirements requires definitions, terms, and equations, as well as the ability to understand how to control their effect and measure their outcomes during system design endeavors.

6.1 Introduction to Conciseness, Modularity, Simplicity and Traceability

This chapter will address four major topics: (1) conciseness; (2) modularity; (3) simplicity or complexity; and (4) traceability in design endeavors. The chapter begins with a section that reviews conciseness and the basic terminology, equations and concepts that underlie its utilization. A metric for measuring and evaluating conciseness is proposed.

Section 6.3 discusses the concept of modularity and how it affects systems designs. A number of specific modularity measures from the extant literature are presented. The section completes with the selection of a measure for modularity and a structural map relating the metric and the measurement attributes for modularity.

Section 6.4 in this chapter addresses simplicity by contrasting it with complexity. Relevant measures for complexity from the related literature are reviewed and three are presented for understanding. The section concludes with a metric and measurable characteristic for complexity.

Section 6.5 presents traceability and how it impacts system design endeavors. The need for traceability expressed in the *IEEE Standard for the Application and Management of the Systems Engineering Process* (IEEE 2005) is used to develop a metric for evaluating traceability in systems designs. The section completes by

K.M. Adams, *Non-functional Requirements in Systems Analysis and Design*,
Topics in Safety, Risk, Reliability and Quality 28,
DOI 10.1007/978-3-319-18344-2_6

relating the proposed measure for traceability as a metric and includes a structural map for traceability.

The chapter has a specific learning goal and associated objectives. The learning goal of this chapter is to be able to identify how the attributes of conciseness, modularity, simplicity and traceability that influence design in systems endeavors. This chapter's goal is supported by the following objectives:

- Define conciseness.
- Describe the terminology used to calculate conciseness.
- Construct a structural map that relates conciseness to a specific metric and measurable characteristic.
- Define modularity.
- Describe the terminology used to represent modularity.
- Construct a structural map that relates modularity to a specific metric and measurable characteristic.
- Describe the relationship between simplicity and complexity.
- Define complexity.
- Describe the terminology used to represent complexity.
- Construct a structural map that relate complexity to a specific metric and measurable characteristic.
- Define traceability.
- Describe the terminology used to represent traceability.
- Construct a structural map that relate traceability to a specific metric and measurable characteristic.
- Explain the significance of conciseness, modularity, simplicity and traceability in systems design endeavors.

The ability to achieve these objectives may be fulfilled by reviewing the materials in the sections that follow.

6.2 Conciseness

In this section the basics of conciseness and how it is applied during systems endeavors is reviewed. Conciseness is not a well-known non-functional requirement and will not have attributes that are either obvious or universally recognized. In order to understand this attribute the review will start with a basic definition.

6.2.1 Conciseness Definitions

Conciseness, from a software engineering perspective, is defined as:

> Software attributes that provide implementation of a function with a minimum amount of code (IEEE and ISO/IEC 2010, p. 69).

This term may be expanded to systems engineering endeavors by utilizing the independence axiom from the theory of Axiomatic Design (Suh 1990, 2001). The *independence axiom*, as utilized in axiomatic design, states:

> Maintain the independence of the functional requirements (Suh 2005, p. 23).

Simply stated, each functional requirement should be satisfied without affecting any other functional requirement.

During the conceptualization process of engineering design, each functional requirement is transformed from the functional domain where they state *what* is required, to the physical domain where they will be matched to a design parameter that will define *how* the function will be accomplished. An ideal, or concise mapping, should be one design parameter for each unique functional requirement. A system that meets this requirement would be ideally concise. Mathematically, this relationship may expressed as the *conciseness ratio* (*CR*), which is expressed in Eq. 6.1.

The Conciseness Ratio

$$CR = \frac{\sum_{i=1}^{n} DP_i}{\sum_{j=1}^{n} FR_j} \tag{6.1}$$

where

i = the number of Design Parameters (*DP*)

j = the number of Functional Requirements (*FR*)

From the definition of conciseness and the associated conciseness ratio (*CR*) it should be clear that designs with a higher conciseness ratio are more concise because their design parameters and functional requirements are *parsimonious*. The next section will address how conciseness is included in the design synthesis process.

6.2.2 Conciseness in Systems Design Efforts

Neither the term conciseness nor the word concise are directly addressed in *IEEE Standard 1220—Systems engineering—Application and management of the systems engineering process* (IEEE 2005). However, requirements analysis (Sect. 6.1) and functional analysis (Sect. 6.3) are serial inputs into the design synthesis process (Sect. 6.5) where design solution alternatives are identified and evaluated.

- As an element of the synthesis task in Sect. 6.5.2 where

> … alternatives and aggregates of alternatives are analyzed to determine which design solution best satisfies allocated functional and performance requirements, interface requirements, and design constraints and adds to the overall effectiveness of the system or higher-level system (IEEE 2005, p. 51).

The conciseness ratio (*CR*) may be applied to design alternatives as a measure for discrimination between competing alternatives.

Table 6.1 Four-level structural map for measuring conciseness

Level	Role
Concern	Systems sustainment
Attribute	Conciseness
Metric	System and component conciseness
Measurable characteristic	*CR*, the conciseness ration which is a measure of the ratio between the number of design parameters (DP) and number of functional requirements (FR) in a system or component

6.2.3 Measuring Conciseness

At the end of Chap. 3 the importance of being able to measure each non-functional attribute was stressed. A structural mapping that relates conciseness to a specific metric and measurement entity are required. The four-level construct for conciseness is presented in Table 6.1.

The section that follows will address the non-functional requirement for modularity as a design concern.

6.3 Modularity

In this section the basics of modularity and how it is applied during systems endeavors will be reviewed. Modularity is a well-established principle in many forms of engineering (Budgen 2003), however, it is not universally applied as a design discriminator. In order to understand this attribute the basic definition serves as a logical starting point.

6.3.1 Modularity Definition

Modularity, from a systems engineering perspective, is defined as:

> 1. The degree to which a system or computer program is composed of discrete components such that a change to one component has minimal impact on other components 2. Software attributes that provide a structure of highly independent components (IEEE and ISO/IEC 2010, p. 223).

Modularity, as a non-functional requirement, has additional definitions, shown in Table 6.2 that may provide improved understanding of the term.

Modularity is typically assessed using two measures: (1) coupling; and (2) cohesion.

Table 6.2 Additional definitions for modularity

Definition	Source
"The property of a system that has been decomposed into a set of cohesive and loosely coupled modules"	Booch (1994, p. 515)
"An encapsulated group of elements that can be manipulated as a unit"	Hornby (2007, p. 52)

6.3.2 Definitions for Coupling and Cohesion

Table 6.3 provides definitions of these terms in a format that permits the reader to easily contrast their differences.

There are a variety of specific types of both cohesion and coupling that are factors to consider during the design of any system. However, systems designers are interested in a higher level of abstraction, that of the system's modularity, and will use the notion of coupling in the development of metrics for modularity.

Table 6.3 Definitions for coupling and cohesion

Coupling	Cohesion
"1. The manner and degree of interdependence between software modules. 2. The strength of the relationships between modules. 3. A measure of how closely connected two routines or modules are. 4. In software design, a measure of the interdependence among modules in a computer program. cf. cohesion. NOTE Types include common-environment coupling, content coupling, control coupling, data coupling, hybrid coupling, and pathological coupling" IEEE and ISO/IEC (2010, p. 83)	"1. The manner and degree to which the tasks performed by a single software module are related to one another. 2. in software design, a measure of the strength of association of the elements within a module. Syn: module strength. cf. coupling. NOTE Types include coincidental, communicational, functional, logical, procedural, sequential, and temporal" IEEE and ISO/IEC (2010, p. 57)
"The degree of interdependence between modules" Yourdon and Constantine (1979, p. 85)	"How tightly bound or related its [i.e., a module] internal elements are" Yourdon and Constantine (1979, p. 106)
"A measure of intermodule connectivity, and is concerned with identifying the forms of connection that exist between modules and the 'strength' of these connections" Budgen (2003, p. 77)	"Provides a measure of the extent to which the components of a module can be considered to be 'functionally related. The ideal module is one in which all of the components can be considered as being solely present for one purpose" Budgen (2003, p. 78)

Table 6.4 Modularity metrics studies by metric type

Metric type	Study reference
Similarity	Newcomb et al. (1998)
	Gershenson et al. (2003, 2004)
Combination	Newcomb et al. (1998)
Coupling	Newcomb et al. (1998)
	Martin and Ishii (2002)
	Mikkola and Gassmann (2003)
	Sosa et al. (2007)
	Hölttä-Otto and de Weck (2007)
	Yu et al. (2007)

6.3.3 Modularity Metrics

A review of the extant literature on modularity metrics reveals 9 unique studies published in scholarly journals. The 9 studies are characterized by the metric type used to develop an appropriate modularity metric. The metric types included the following metric types: (1) similarity, (2) coupling, and (3) a combination of similarity and coupling and are presented in Table 6.4.

While and in-depth review of each study is beyond the scope of this chapter, readers are encouraged to consult the references for a more detailed description of the development of each modularity metric. The sections that follow will offer a brief glimpse into two of the more robust modularity metrics.

6.3.3.1 The Modularization Function

The study by Mikkola and Gassmann (2003) introduces a mathematical model, termed the modularization function, that is used to analyze the degree of modularity in a given product architecture. The modularization function depends on three variables: (1) components (i.e., modules); (2) degree of module coupling; and (3) substitutability of New-To-the-Firm (NTF) components (i.e., modules). The modularization function is presented in Eq. 6.2.

Modularization Function

$$M(u) = e^{-u^2/2Ns\delta} \tag{6.2}$$

where:

$M(u)$ the modularization function
u the number of New-To-Firm (NTF) components (i.e., modules)
N total number of components (i.e., modules)
s substitutability factor
δ degree of module coupling

6.3.3.2 Minimum Description Length (MDL)

The Design Structure Matrix or DSM (Browning 2001; Eppinger and Browning 2012; Steward 1981) is a widely used design approach that has, as its roots, both the simplified N^2 diagram (Becker et al. 2000) and the House of Quality (Hauser and Clausing 1988). The DSM is a graphical method for representing the relationships between the components (i.e., modules) of a system. The DSM is a square matrix with identical labels for both the rows and columns. In a static, component-based DSM, the labels represent the system architecture by depicting the relevant relationships between the system's constituent components or modules.

In order to illustrate a basic DSM, a simple refrigeration system is depicted in Fig. 6.1.

Table 6.5 is a sample DSM for the simple refrigeration system depicted in Fig. 6.1.

The type of interaction that occurs between the system components or modules is characterized in Table 6.6.

The relationships between the components are coded based upon (1) the component or module interaction type as described in Table 6.6, and (2) weighting the interactions between components or modules relative to each other as presented in Table 6.7.

The DSM structure for the system in Fig. 6.1 can be used to display any of the interaction Types from Table 6.6. A completed DSM for the Energy interactions, using the weighting schema from Table 6.7 is presented in Table 6.8.

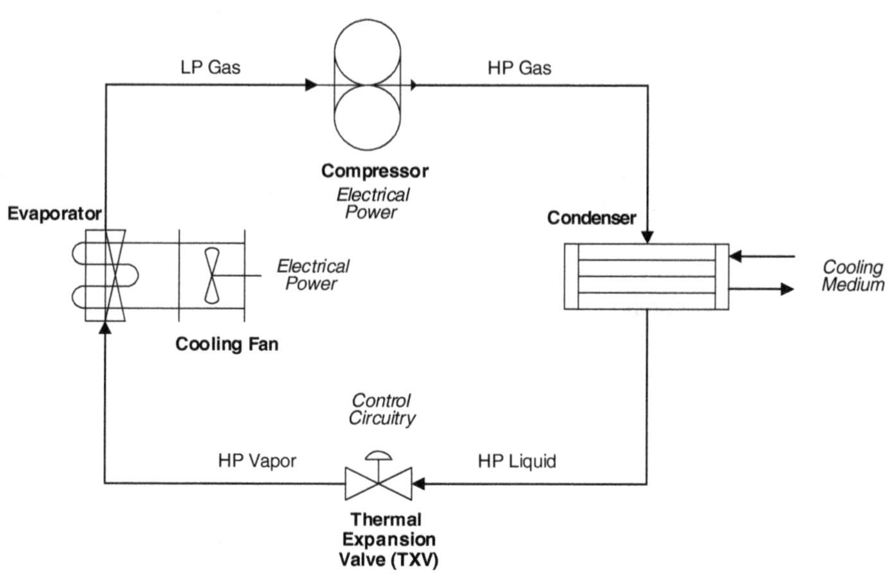

Fig. 6.1 Simplified single-stage vapor compression refrigeration system

Table 6.5 DSM structure for system depicted in Fig. 6.1

		A	B	C	D	E	F	G	H
Compressor	A	A							
Condenser	B		B						
TXV	C			C					
Evaporator	D				D				
Cooling fan	E					E			
Electrical power	F						F		
Cooling medium	G							G	
Control circuitry	H								H

Table 6.6 Module interaction types (Browning 2001, p. 294)

Interaction type	Description
Spatial	Association of physical space and alignment, needs for adjacency or orientation between two elements
Energy	The need for energy transfer/exchange between two elements (e.g., power supply)
Information	Need for data or signal exchange between two elements
Material	Need for material exchange between two elements

The study by Yu et al. (2007) proposes a modularity metric termed the Minimum Description Length (MDL), that is based upon an evaluation of a design as represented in a DSM. The MDL evaluates the system's modularity (represented by the DSM) using the terms presented in Eq. 6.3.

Minimum Description Length

$$MDL = \frac{1}{3}\left(n_c \log n_n + \log n_n \sum_{i=1}^{n_c} cl_i\right) + \frac{1}{3}S_1 + \frac{1}{3}S_2 \quad (6.3)$$

where

n_c the number of components (i.e., modules)
n_n the number of rows or columns in the DSM
cl_i the size of the module i
S_1 the number of cells that are in a module, but are empty
S_2 the number of cells that is 1 in between the modules

There are a number of methods for determining the degree of modularity in a complex system. In this section, two methods have been presented that address modularity as a function of coupling that may serve as useful measures of modularity.

Table 6.7 Weighting schema for energy interaction between components (Browning 2001)

Label	Weight	Description
Required	+2	Energy transfer/exchange is necessary for functionality
Desired	+1	Energy transfer/exchange is beneficial, but not necessary for functionality
Indifferent	0	Energy transfer/exchange does not affect functionality
Undesired	−1	Energy transfer/exchange causes negative effects but does not prevent functionality
Detrimental	−2	Energy transfer/exchange must be prevented to achieve functionality

Table 6.8 DSM for energy actions in the system depicted in Fig. 6.1

		A	B	C	D	E	F	G	H
Compressor	A	A					+2		
Condenser	B		B					+2	
TXV	C			C					+2
Evaporator	D				D				
Cooling fan	E					E			+2
Electrical power	F	+2		+2		+2	F		+2
Cooling medium	G		+2					G	
Control circuitry	H		+2			+2			H

6.3.4 Modularity in Systems Design Efforts

Neither the term modularity nor the word coupling are directly addressed in *IEEE Standard 1220—Systems engineering—Application and management of the systems engineering process* (IEEE 2005). However, the reason that modularity is a feature in the purposeful design of systems is because, from an engineering design perspective modularity does many things:

- First, it makes the complexity of the system manageable by providing an effective "division of cognitive labor."
- Second, modularity organizes and enable parallel work.
- Finally, modularity in the 'design' of a complex system allows modules to be changed and improved over time without undercutting the functionality of the system as a whole (Baldwin and Clark 2006, p. 180).

However, modularity-in-design is not defined simply as a system with a defined hierarchy of modules. "A complex engineering system is modular in design if (and only if) the *process of designing it can be split up and distributed across separate modules*, that are coordinated by design rules, not by ongoing consultations amongst the designers" (Baldwin and Clark 2006, p. 181).

Table 6.9 Four-level structural map for measuring modularity

Level	Role
Concern	Systems design
Attribute	Modularity
Metric	System modularity
Measurable characteristic	Number of modules and degree of coupling as represented by either the modularization function (M_u) or the Minimum Description Length (*MDL*)

The presence of design rules is mandatory for modularity-in-design. Having design rules permits the system and module designers to create a wide variety of design options that may be considered as elements of the overall system solution. The design rules hold the system designers accountable for basic elements of design, but do not overly constrain the potential solutions and independent options possible for each module. The essence of this approach is that modularization permits the final system to be changed and improved over its life cycle without disabling the overall functionality of the system.

6.3.5 Measuring Modularity

At the end of Chap. 3 the importance of being able to measure each non-functional attribute was stressed as an important element of the design process. A structural mapping that relates modularity to a specific metric and measurement entity are required. The four-level construct for modularity is presented in Table 6.9.

The section that follows will address the non-functional requirement for simplicity as a design concern.

6.4 Simplicity

In this the basics of simplicity and how it is applied during systems endeavors will be reviewed. "While simplicity cannot be assessed, one can at least seek measures for its converse characteristic of *complexity*" (Budgen 2003, p. 75). In order to understand simplicity and complexity it is best to once again start with their basic definitions.

6.4.1 Simplicity and Complexity Definitions

Simplicity and its converse characteristic complexity are defined, from a systems engineering perspective, in Table 6.10. The definitions are provided side-by-side to permit the reader to easily contrast their differences.

Table 6.10 Definitions for simplicity and complexity

Simplicity	Complexity
"1. The degree to which a system or component has a design and implementation that is straightforward and easy to understand 2. Software attributes that provide implementation of functions in the most understandable manner cf. complexity" IEEE and ISO/IEC (2010, p. 327)	"1. The degree to which a system's design or code is difficult to understand because of numerous components or relationships among components. 2. Pertaining to any of a set of structure-based metrics that measure the attribute in (1). 3. The degree to which a system or component has a design or implementation that is difficult to understand and verify cf. simplicity" IEEE and ISO/IEC (2010, p. 63)

Because simplicity cannot be directly measured it will be measured as the inverse of complexity, which has some basic measures. However, before complexity may be measured, additional understanding of the term and how it may be characterized is in order.

6.4.2 Characteristics of Complexity

The definition for complexity in Table 6.10 does not provide sufficient level of detail required to understand what complexity is and how it appears in systems. It is often useful to characterize complexity by the features that would be present in a system that is characterized as complex. Typically, these include:

1. The system contains a collection of many interacting objects or agents.
2. These objects' behavior is affected by memory or feedback.
3. The objects can adapt their strategies according to their history.
4. The system is typically open.
5. The system appears to be alive.
6. The system exhibits phenomena which are generally surprising, and may be extreme.
7. The emergent phenomena typically arise in the absence of any sort of invisible hand or central controller.
8. The system shows a complicated mix of ordered and disordered behavior (Johnson 2007, pp. 13–15).

Armed with this improved understanding of complexity, how is it measured? The next section will address methods for measuring complexity in systems.

6.4.3 Methods for Measuring Complexity in Systems

Detailed methods for measuring complexity in software and generic non-software systems have been addressed in the literature. Some of the prominent studies are listed in Table 6.11.

Table 6.11 Literature on measuring complexity in systems

Systems type	Complexity measurement approach	References
Software	Control flows	McCabe (1976), McCabe and Butler (1989)
	Information flows	Henry and Kafura (1984), Kitchenham et al. (1990)
	Comprehension	Halstead (1977)
	Elements	Briand et al. (2000), Chidamber and Kemerer (1994)
Generic systems	Hierarchical structures	Huberman and Hogg (1986)
	Design effort	Bashir and Thomson (1999)
	Design problem, process and product	Ameri et al. (2008), Summers and Shah (2010)
	Structural and functional complexity in design	Braha and Maimon (1998)
	Human performance	Henneman and Rouse (1986)
	Information theory and entropy	Conant (1976), Jung and Cho (1996), Koomen (1985), Min and Soon Heung (1991)
	Variety	Ashby (1958, 1968), Bar-Yam (2004)

While an in-depth review of each approach to measuring complexity is beyond this chapter, readers are encouraged to consult the references for a more detailed description of the development of each complexity measurement approach. The sections that follow will offer a brief glimpse into three of the more robust and generalizable complexity metrics approaches for system design endeavors.

6.4.3.1 System Hierarchy Tree

One measure for system complexity that is particular useful during system design has been proposed by Huberman and Hogg (1986). The authors report that their physical measure for system complexity is based on "its diversity, while ignoring its detailed specification. It applies to discrete hierarchical structures made up of elementary parts and provides a precise, readily computable quantitative measure" (p. 376). Their method relies upon the concept of hierarchy, and utilizes a hierarchy tree to represent the structure of a system.

> A powerful concept for understanding these systems is that of a hierarchy. This can correspond to the structural layout of a system or, more generally, to clustering pieces by strength of interaction. In particular, if the most strongly interacting components are grouped together at the first level, then the most strongly interacting clusters are combined at the next level and so on, one ends up with a tree reflecting the resulting hierarchy (Huberman and Hogg 1986, p. 377).

The structure for a notional system is depicted in the hierarchy tree in Fig. 6.2.

The premise of hierarchy upon which this measure relies is that proposed by Simon (1996).

> To design such a complex structure, one powerful technique is to discover viable ways of decomposing it into semi-independent components corresponding to its many functional parts. The design of each component can then be carried out with some degree of independence of the design of others, since each will affect the others largely through its function and independently of the details of the mechanisms that accomplish the function (p. 128).

The complexity of the system *C(T)*, is a function of its tree-like hierarchical structure and is related by Diversity *D(T)*, in Eqs. 6.4 and 6.5.

Complexity as a Function of Diversity

$$C(T) = 1 - D(T) \tag{6.4}$$

Complexity in a Hierarchical Tree

$$D(T) = (2^k - 1)\prod_{j=1}^{k} D(T_{i_j}) \tag{6.5}$$

where

C complexity measure
D diversity measure
T system hierarchy tree whose diversity is being evaluated
J number of non-isomorphic sub-trees with a range from 1 to *k*
k number of sub-trees in the system elements

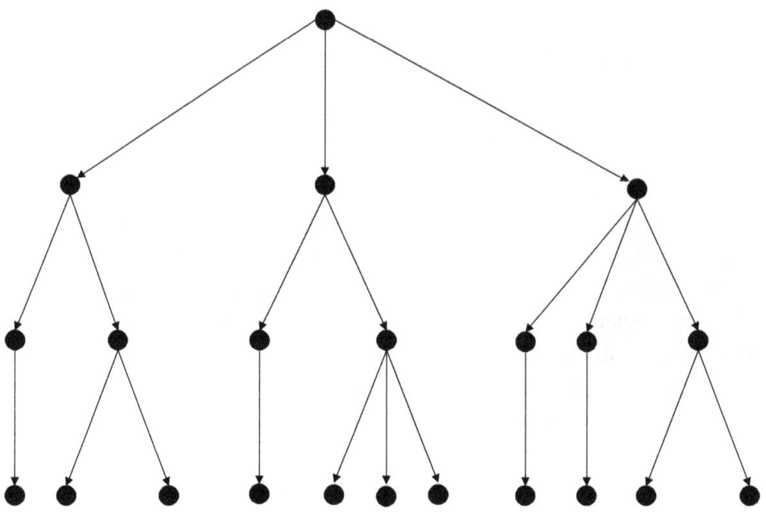

Fig. 6.2 System structure depicted in a hierarchy tree

6.4.3.2 Information Theory and Entropy

A second measure for system complexity is based upon information theory. Information theory has, at its roots, the groundbreaking work conducted by Shannon (1948a, b; Shannon and Weaver 1998) on communications that related entropy and information. Claude Shannon [1916–2001], the father of Information Theory, adapted these concepts to the analysis of entropy in information, stating:

> That information be measured by entropy is, after all, natural when we remember that information, in communication theory, is associated with the amount of freedom of choice we have in constructing a message (Shannon and Weaver 1998, p. 13).

Shannon's conceptualization of information entropy is defined in Eq. 6.6.

Shannon's Information Entropy

$$H = -\sum_{i} p_i \log_2 p_i \tag{6.6}$$

where
H is the information entropy
B is the base 2 logarithm due to the use of binary logic in information theory
p is the probability associated with each the symbols
i the number of discrete messages

During the discussion of the Axiomatic Design Methodology in Chap. 2, Suh (1990, 2001) reformulated Shannon's information entropy in Eq. 6.6 such that the reformulated equation for information content (I), as related to the probability (p) of a design parameter (DP_i) satisfying a functional requirement (FR_i) is presented in Eq. 6.7

System Information Content

$$I_{sys} = -\sum_{i=1}^{n} \log_2 [p(DP_i)] \tag{6.7}$$

The information axiom, when used in this context, states that the system design with the smallest I_{sys} (i.e., the design with the least amount of information) is the best design. This is because this design requires the least amount of information to fulfill the design parameters (DP). Once again, the Axiomatic Design Methodology's utilization of Shannon's information entropy is remarkable because a system's design complexity, most often expressed as a qualitative assessment, may be represented as a quantitative measure based on the information entropy required to satisfy the design parameters.

6.4.3.3 System Variety

None of the previous measures are easily employed as a truly generalizable measure of a system's complexity. However, there is a highly generalizable measure termed *variety* that may be used as a measure of the complexity of a system. Variety is "the total number of possible states of a system, or of an element of a system" (Beer 1981, p. 307). As such, it is an excellent measure of the complexity of a system.

Variety, as a measure of system complexity, computes the number of different possible system states that may exist and is calculated using the relations in Eq. 6.8 (Flood and Carson 1993, p. 26).

Variety

$$V = Z^n \tag{6.8}$$

where
V variety or potential number of system states
Z number of possible states of each system element
N number of system elements

The variety of a simple system can quickly become enormous. Take, for example, a system that has 8 different subsystems with 8 possible channels capable of operating simultaneously. This is a total of 64 different system elements. In this system each element can only have 2 element states—working or not working. The variety generated from the system show that the system may have 18,446,744,073,710,000,000 states!

Of the three methods for measuring complexity, variety seems to be the easiest to compute based upon a minimum of required characteristics for its calculation.

6.4.4 Measuring Complexity

As stated in each of the previous sections, the importance of being able to measure each non-functional attribute is essential in systems design endeavors. A structural mapping that relates complexity to a specific metric and measurement entity are required. The four-level construct for complexity is presented in Table 6.12.

Table 6.12 Four-level structural map for measuring complexity

Level	Role
Concern	Systems design
Attribute	Complexity
Metric	System complexity
Measurable characteristic	Number of system modules or components and the possible states of each element or module (i.e., variety)

The section that follows will address the non-functional requirement for traceability as a design concern.

6.5 Traceability

In this section the basics of traceability and how it is applied during systems endeavors will be reviewed. In order to understand traceability its definition is reviewed.

6.5.1 Traceability Definitions

Traceability, from a systems engineering perspective, is defined as

> 1. The degree to which a relationship can be established between two or more products of the development process, especially products having a predecessor-successor or master-subordinate relationship to one another. 2. The identification and documentation of derivation paths (upward) and allocation or flowdown paths (downward) of work products in the work product hierarchy (IEEE and ISO/IEC 2010, p. 378).

Traceability, as a non-functional requirement, has additional definitions in Table 6.13 that may provide improved understanding of the term.

Based upon these definitions traceability is a non-functional requirements that can be viewed from within the life cycle of the system with which it is associated. As such, the concept of traceability starts with the system's stakeholders, as depicted in Fig. 6.3.

The concept depicted in Fig. 6.3 conforms to the traceability process proposed by Jarke (1998) that has four traceability links:

- **Forward from requirements**. Responsibility for requirements achievement must be assigned to system components, such that accountability is established and the impact of requirements change can be evaluated.
- **Backward to requirements**. Compliance of the system with requirements must be verified, and gold-plating (designs for which no requirements exist) must be avoided.

Table 6.13 Definitions for traceability

Definition	Source
"The process of defining logical links between one system element (use case, functional requirements, business rule, design component, code module, test case, and the like) and another"	Wiegers (2003, p. 490)
"A discernable association among two or more logical entities such as requirements, systems elements, verifications, or tasks"	Chrissis et al. (2007, p. 636)

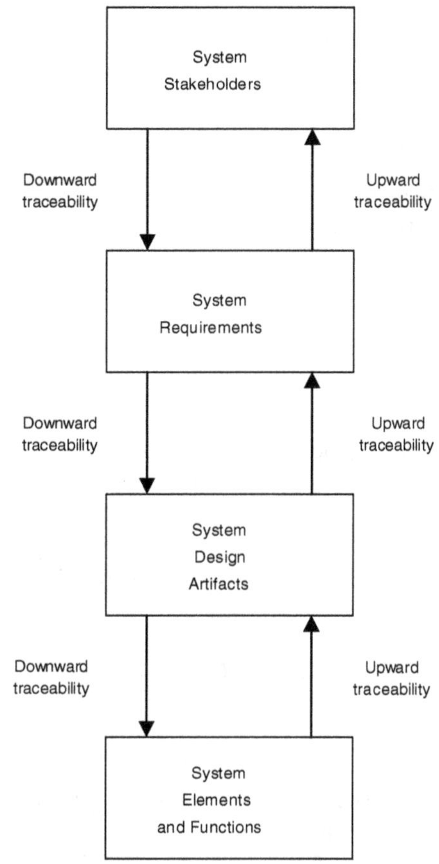

Fig. 6.3 Design traceability in the systems life cycle

- **Forward to requirements**. Changes in stakeholder needs, as well as in technical assumptions, may require a radical reassessment of requirements relevance.
- **Backward from requirements**. The contribution structures underlying requirements are crucial in validating requirements, especially in highly political settings (p. 32).

The ability to trace stakeholder requirements to systems requirements and system requirements to design artifacts and from design artifacts to systems elements and functions (i.e., forward traceability) and the reverse (backward traceability) is an essential feature of all properly documented system design endeavors. Forward traceability ensures that the system is built to satisfy requirements specified and endorsed by the stakeholders. Backward traceability ensures that requirements not formally endorsed by the customer have not *crept* into the system design in what is known as *requirements creep*.

> Requirements creep must be guarded against as additional unendorsed requirements ultimately compete with endorsed requirements for valuable design space and may end up reducing the performance of the system against the endorsed requirements. Unendorsed requirements may also add significantly to the costs and schedule associated with the system, development, without adding to the system's ability to solve the original stakeholder need (Faulconbridge and Ryan 2003, p. 41).

Now that traceability has been formally defined, how it is instantiated as part of the formal design process may be reviewed.

6.5.2 Traceability in Systems Design Efforts

It is important to note that many important questions about the design of a system can only be answered by understanding the relationships between the design layers depicted in Fig. 6.3. "Documenting these relationships engenders greater reflection and subjects your thinking to peer review" (Dick 2005, p. 14). The formal design process is where the traceability relationships are developed.

Traceability, is a major factor in ensuring a robust system design that satisfies the stakeholder's identified needs. Traceability is addressed in *IEEE Standard 1220—Systems engineering—Application and management of the systems engineering process* (IEEE, 2005) which specifically addresses traceability in nine specific areas.

1. As an element of the system definition stage described in Section 5.11.3.

   > System product functional and performance requirements should be allocated among the subsystems so as to assure requirement traceability from the system products to their respective subsystems, and from subsystems to their parent product (IEEE 2005, p. 22).

2. As an element of the preliminary design stage described in Section 5.2.1.1.

   > Subsystem performance requirements are allocated among the assemblies so as to assure requirements traceability from subsystems to appropriate assemblies, and from assemblies to the parent subsystem (IEEE 2005, p. 25).

3. As an element of the preliminary design stage described in Section 5.2.1.2.

   > Assembly performance requirements are allocated among the components so as to assure requirement traceability from the assemblies to their respective components, and from components to their parent assembly (IEEE 2005, p. 25).

4. As an element of the detailed design stage described in Section 5.3.1.1.

   > Component requirements should be allocated among the subcomponents in a manner that ensures that requirements traceability is maintained in both directions (IEEE 2005, p. 29).

5. As an element of the functional analysis process described in Section 6.3.1.3.

   > The project documents the allocation of system performance requirements to functions to provide traceability and to facilitate later changes (IEEE 2005, p. 46).

6. As an element of the design synthesis process described in Sect. 6.5.1.

> Requirements traceability is established and recorded to ensure that all functions are allocated to elements of the system; each system element performs at least one function (IEEE 2005, p. 51).

7. As an element of the design synthesis process described in Section 6.5.18.

> The design architecture includes the requirements traceability and allocation matrices, which capture the allocation of functional and performance requirements among the system elements (IEEE 2005, p. 53).

8. As an element of the design verification process described in Section 6.6.2.1.

> Design elements descriptions are traceable to requirements of the functional architecture (upward traceability) (IEEE 2005, p. 55).

9. As an element of the design verification process described in Section 6.6.8.

> The project shall generate an SBS to depict the hierarchy of products and processes that comprise the system architecture. The SBS can be used by the project for … traceability (IEEE 2005, p. 17).

6.5.3 A Method for Evaluating Traceability

Now traceability has been defined within the processes used to design of systems, how is it measured? This question is a tough one to answer because traceability is a subjective, qualitative measure which differs from the objective, quantitative measures developed for most of the non-functional requirements already addressed. In order to understand how to approach a subjective, qualitative measure, a short review of how to measure subjective, qualitative objects is required.

6.5.3.1 Development of Measurement Scales

In order to evaluate traceability, questions that address both the presence (yes or no) and quality of the effort (how well) to provide traceability in the nine areas where traceability is addressed in systems design. In this case each of the nine areas or objects must be related to a specific measurable attribute. Measures are important because they are the linkage between observable, real-world, empirical facts and the construct (i.e., traceability) that are created as an evaluation point. In this case a measure is defined as "… an observed score gathered through self-report, interview, observation, or some other means" (Edwards and Bagozzi 2000, p. 156).

The selection of a measurement scale is an important element in the development of an adequate measure for traceability. A scale is defined as "… a theoretical variable in a model, and *scaling* or measurement is the attachment to empirical events of values of the variable in a model" (Cliff 1993, p. 89). Because measurement "... concerns the assignment of numbers to objects to represent amounts or degrees of a property possessed by all of the objects," (Torgerson 1958, p. 19) the type of scale is an important selection criteria. Stevens (1946), in his seminal work on measurement states that "the type of scale achieved depends upon the character of the basic empirical operations performed. These operations are limited ordinarily by the nature of the thing being scaled and by our choice of procedures" (p. 677). Scale selection falls into one or another of the four scale types (Coombs et al. 1954). Because the nine measurement areas or objects we have selected for traceability have no natural origin or empirically defined distance, the ordinal scale was selected as an appropriate scale for measuring the traceability attributes. Ordinal scales are those in which three criteria have been met:

> (1) A set of objects is ordered from most to least with respect to an attribute, (2) there is no indication of how much in an absolute sense any of the objects possess the attribute, and (3) there is no indication of how far apart the objects are with respect to the attribute (Nunnally 1967, p. 12).

The numbers attached to the ordinal scale only provide a shorthand notation for designating the relative positions of the measures on the scale. The use of a well-known scale type, the Likert scale, is proposed for use in evaluating traceability. Because Likert-type ordinal scales have been shown to have increased reliability (as measured by Cronbach's (1951) Coefficient alpha) up to the use of 5 points and "... a definite leveling off in the increase in reliability after 5 scale points," (Lissitz and Green 1975, p. 13) the scales used for traceability in the next section have been purposefully designed for increased reliability by using 5 points.

Before moving on to describing the measure for traceability two important points must be made with respect to scale development. Scales are characterized as either a *proposed* scale or a scale. "A proposed scale is one that some investigator(s) put forward as having the requisite properties, and if it is indeed shown to have them, then it is recognized as a scale" (Cliff 1993, p. 65). In this chapter the use of the word scale is referring to *proposed scales*. This may seem to be an insignificant point, but until the scale has been accepted and successfully utilized it remains proposed. The second and final point is that the use of an ordinal scale limits the measurement effort to but a few statistics such as "... rank order coefficient of correlation, *r*, Kendall's *W*, and rank order analysis of variance, medians, and percentiles" (Kerlinger and Lee 2000, p. 363). Because the current evaluation techniques for traceability use few if any measures, the statistical limitation imposed by the use of ordinal scales may be evaluated as an acceptable one.

6.5.3.2 Proposed Measurement Scale for Traceability

Armed with a construct, measurement attributes and an appropriate scale type, the traceability measure may be constructed. In order to evaluate traceability, we will need to answer questions that address both the presence (yes or no) and quality of the effort (how well) to provide traceability in the nine areas (i.e., our measurement

Table 6.14 Measurement constructs for traceability

Life cycle stage or process	IEEE std 1220 section	Traceability concern for measurement
Conceptual design	5.1.13	System product functional and performance requirements should be allocated among the subsystems so as to assure requirement traceability from the system products to their respective subsystems, and from subsystems to their parent product
Preliminary design	5.2.1.1	Subsystem performance requirements are allocated among the assemblies so as to assure requirements traceability from subsystems to appropriate assemblies, and from assemblies to the parent subsystem
	5.2.1.2	Assembly performance requirements are allocated among the components so as to assure requirement traceability from the assemblies to their respective components, and from components to their parent assembly
Detailed design	5.3.1.1	Component requirements should be allocated among the subcomponents in a manner that ensures that requirements traceability is maintained in both directions
Functional analysis	6.3.1	The project documents the allocation of system performance requirements to functions to provide traceability and to facilitate later changes
Design synthesis	6.5.1	Requirements traceability is established and recorded to ensure that all functions are allocated to elements of the system; each system element performs at least one function
	6.5.18	The design architecture includes the requirements traceability and allocation matrices, which capture the allocation of functional and performance requirements among the system elements
Design verification	6.6.2.1	Design elements descriptions are traceable to requirements of the functional architecture (upward traceability)
	6.6.8	The project shall generate an SBS to depict the hierarchy of products and processes that comprise the system architecture. The SBS can be used by the project for ... traceability

constructs) from Section 5.4.2. Table 6.14 has rearranged the nine constructs and associated measurement concerns based upon the life cycle stage and systems engineering process.

In order to evaluate the design's ability to conform to the notion of traceability, a specific question should be developed which will evaluate each of the nine design traceability measurement concerns. The answers to the questions will be contained in a 5 point-Likert scale. The measurement constructs and questions associated with each of the measurements concerns are presented in Table 6.15.

The answer to each question in Table 6.15 will be scored using the 5-point Likert measures in Table 6.16.

Table 6.15 Measurement questions for design traceability

Measurement construct	Traceability concern for measurement
T_{cd}	Does the conceptual design provide objective quality evidence for requirement traceability from the system products to their respective subsystems, and from subsystems to their parent product?
T_{pd1}	Does the preliminary design provide objective quality evidence for traceability from subsystems to appropriate assemblies, and from assemblies to the parent subsystem?
T_{pd2}	Does the preliminary design provide objective quality evidence for traceability from the assemblies to their respective components, and from components to their parent assembly?
T_{dd}	Does the detailed design provide objective quality evidence that ensures that requirements traceability from component to subcomponent is maintained in both directions?
T_{fa}	Does the functional analysis process provide objective quality evidence for the allocation of system performance requirements to functions to provide traceability and to facilitate later changes?
T_{s1}	Does the design synthesis process provide objective quality evidence to ensure that requirements traceability is established and recorded to ensure that all functions are allocated to elements of the system and that each system element performs at least one function?
T_{s2}	Does the design synthesis process provide objective quality evidence in the design architecture includes the requirements traceability and allocation matrices, which capture the allocation of functional and performance requirements among the system elements?
T_{v1}	Does the verification process provide objective quality evidence that demonstrates design elements descriptions are traceable to requirements of the functional architecture (upward traceability)?
T_{v2}	Does the verification process provide objective quality evidence that a system breakdown structure has been generated to depict the hierarchy of products and processes that comprise the system architecture?

Table 6.16 Traceability measurement question Likert scale

Measure	Descriptor	Measurement criteria
0.0	None	No objective quality evidence is present
0.5	Limited	Limited objective quality evidence is present
1.0	Nominal	Nominal objective quality evidence is present
1.5	Wide	Wide objective quality evidence is present
2.0	Extensive	Extensive objective quality evidence is present

Table 6.17 Four-level structural map for measuring traceability

Level	Role
Concern	Systems design
Attribute	Traceability
Metric	System traceability
Measurable characteristic	Traceability of (1) conceptual design (T_{cd}), (2) preliminary design (T_{pd1}, T_{pd2}), (3) detailed design (T_{dd}), (4) functional analysis (T_{fa}), (5) design synthesis (T_{s1}, T_{s2}), and (6) verification (T_{v1}, T_{v2})

The overall measure for system traceability is a sum of the scores from the nine traceability metrics as shown in Eq. 6.9.

Generalized Equation for System Traceability

$$T_{sys} = \sum_{i=1}^{n} T_i \tag{6.9}$$

Expanded Equation for System Traceability

$$T_{sys} = T_{cd} + T_{pd1} + T_{pd2} + T_{dd} + T_{fa} + T_{s1} + T_{s2} + T_{v1} + T_{v2} \tag{6.10}$$

The summation of the nine (9) constructs in Eq. 6.10 will be the measure the degree of traceability in a system design endeavor Table 6.16.

6.5.4 Measuring Traceability

As stated in each of the three previous sections, the importance of being able to measure each non-functional attribute is essential in systems design endeavors. A structural mapping that relates traceability to a specific metric and measurement entity are required. The four-level construct for traceability is presented in Table 6.17.

6.6 Summary

In this chapter the non-functional requirements for conciseness, modularity, simplicity, and traceability have been reviewed. In each case a formal definition has been provided along with additional explanatory definitions, terms, and equations. The ability to effect the non-functional requirement during the design process has also been addressed. Finally, a formal metric and measurement characteristic have been proposed for evaluating each non-functional requirement attribute.

The chapter that follows will address non-functional requirement for compatibility, consistency, interoperability, and safety as part of the concern for design in systems endeavors.

References

Ameri, F., Summers, J. D., Mocko, G. M., & Porter, M. (2008). Engineering design complexity: An investigation of methods and measures. *Research in Engineering Design, 19*(2–3), 161–179.

Ashby, W. R. (1958). Requisite variety and its implications for the control of complex systems. *Cybernetica, 1*(2), 83–99.

Ashby, W. R. (1968). Variety, constraint, and the law of requisite variety. In W. Buckley (Ed.), *Modern systems research for the behavioral scientist* (pp. 129–136). Chicago: Aldine Publishing Company.

Baldwin, C. Y., & Clark, K. B. (2006). Modularity in the design of complex engineering systems. In D. Braha, A. A. Minai, & Y. Bar-Yam (Eds.), *Complex engineered systems* (pp. 175–205). Berlin: Springer.

Bar-Yam, Y. (2004). Multiscale variety in complex systems. *Complexity, 9*(4), 37–45.

Bashir, H. A., & Thomson, V. (1999). Estimating design complexity. *Journal of Engineering Design, 10*(3), 247–257.

Becker, O., Asher, J. B., & Ackerman, I. (2000). A method for system interface reduction using N2 charts. *Systems Engineering, 3*(1), 27–37.

Beer, S. (1981). *Brain of the Firm.* New York: Wiley.

Booch, G. (1994). *Object-oriented analysis and design with applications* (2nd ed.). Reading, MA: Addison-Wesley.

Braha, D., & Maimon, O. (1998). The measurement of a design structural and functional complexity. *IEEE Transactions on Systems, Man and Cybernetics—Part A: Systems and Humans, 28*(4), 527–535.

Briand, L. C., Wüst, J., Daly, J. W., & Victor Porter, D. (2000). Exploring the relationships between design measures and software quality in object-oriented systems. *Journal of Systems and Software, 51*(3), 245–273.

Browning, T. R. (2001). Applying the design structure matrix to system decomposition and integration problems: A review and new directions. *IEEE Transactions on Engineering Management, 48*(3), 292–306.

Budgen, D. (2003). *Software design* (2nd ed.). New York: Pearson Education.

Chidamber, S. R., & Kemerer, C. F. (1994). A metrics suite for object oriented design. *IEEE Transactions on Software Engineering, 20*(6), 476–493.

Chrissis, M. B., Konrad, M., & Shrum, S. (2007). *CMMI: Guidelines for process integration and product improvement* (2nd ed.). Upper Saddle River, NJ: Addison-Wesley.

Cliff, N. (1993). What is and isn't measurement. In G. Keren & C. Lewis (Eds.), *A handbook for data analysis in the behavioral sciences: Methodological issues* (pp. 59–93). Hillsdale, NJ: Lawrence Erlbaum Associates.

Conant, R. C. (1976). Laws of information which govern systems. *IEEE Transactions on Systems, Man and Cybernetics, SMC, 6*(4), 240–255.

Coombs, C. H., Raiffa, H., & Thrall, R. M. (1954). Some views on mathematical models and measurement theory. *Psychological Review, 61*(2), 132–144.

Cronbach, L. J. (1951). Coefficient alpha and the internal structure of tests. *Psychometrika, 16*(3), 297–334.

Dick, J. (2005). Design traceability. *IEEE Software, 22*(6), 14–16.

Edwards, J. R., & Bagozzi, R. P. (2000). On the nature and direction of relationships between constructs and measures. *Psychological Methods, 5*(2), 155–174.

Eppinger, S. D., & Browning, T. R. (2012). *Design structure matrix methods and applications*. Cambridge, MA: MIT Press.

Faulconbridge, R. I., & Ryan, M. J. (2003). *Managing complex technical projects: A systems engineering approach*. Norwood, MA: Artech House.

Flood, R. L., & Carson, E. R. (1993). *Dealing with complexity: An introduction to the theory and application of systems science* (2nd ed.). New York: Plenum Press.

Gershenson, J. K., Prasad, G. J., & Zhang, Y. (2003). Product modularity: Definitions and benefits. *Journal of Engineering Design, 14*(3), 295.

Gershenson, J. K., Prasad, G. J., & Zhang, Y. (2004). Product modularity: Measures and design methods. *Journal of Engineering Design, 15*(1), 33–51.

Halstead, M. H. (1977). *Elements of Software Science*. Elsevier North-Holland, Inc., Amsterdam.

Hauser, J. R., & Clausing, D. P. (1988). The house of quality. *Harvard Business Review, 66*(3), 63–73.

Henneman, R. L., & Rouse, W. B. (1986). On measuring the complexity of monitoring and controlling large-scale systems. *IEEE Transactions on Systems, Man and Cybernetics, 16*(2), 193–207.

Henry, S., & Kafura, D. (1984). The evaluation of software systems' structure using quantitative software metrics. *Software: Practice and Experience, 14*(6), 561–573.

Hölttä-Otto, K., & de Weck, O. (2007). Degree of modularity in engineering systems and products with technical and business constraints. *Concurrent Engineering, 15*(2), 113–126.

Hornby, G. S. (2007). Modularity, reuse, and hierarchy: Measuring complexity by measuring structure and organization. *Complexity, 13*(2), 50–61.

Huberman, B. A., & Hogg, T. (1986). Complexity and adaptation. *Physica D: Nonlinear Phenomena, 22*(1–3), 376–384.

IEEE. (2005). *IEEE standard 1220: Systems engineering—application and management of the systems engineering process*. New York: Institute of Electrical and Electronics Engineers.

IEEE, & ISO/IEC (2010). *IEEE and ISO/IEC standard 24765: Systems and software engineering—vocabulary*. New York and Geneva: Institute of Electrical and Electronics Engineers and the International Organization for Standardization and the International Electrotechnical Commission.

Jarke, M. (1998). Requirements tracing. *Communications of the ACM, 41*(12), 32–36.

Johnson, N. (2007). *Simply complexity: A clear guide to complexity theory*. Oxford: Oneworld Publications.

Jung, W. S., & Cho, N. Z. (1996). Complexity measures of large systems and their efficient algorithm based on the disjoint cut set method. *IEEE Transactions on Nuclear Science, 43*(4), 2365–2372.

Kerlinger, F. N., & Lee, H. B. (2000). *Foundations of behavioral research*. Fort Worth: Harcourt College Publishers.

Kitchenham, B. A., Pickard, L. M., & Linkman, S. J. (1990). An evaluation of some design metrics. *Software Engineering Journal, 5*(1), 50–58.

Koomen, C. J. (1985). The entropy of design: A study on the meaning of creativity. *IEEE Transactions on Systems, Man and Cybernetics, SMC, 15*(1), 16–30.

Lissitz, R. W., & Green, S. B. (1975). Effect of the number of scale points on reliability: A Monte Carlo approach. *Journal of Applied Psychology, 60*(1), 10–13.

Martin, M. V., & Ishii, K. (2002). Design for variety: Developing standardized and modularized product platform architectures. *Research in Engineering Design, 13*(4), 213–235.

McCabe, T. J. (1976). A complexity measure. *IEEE Transactions on Software Engineering, SE, 2*(4), 308–320.

McCabe, T. J., & Butler, C. W. (1989). Design complexity measurement and testing. *Communications of the ACM, 32*(12), 1415–1425.

Mikkola, J. H., & Gassmann, O. (2003). Managing modularity of product architectures: Toward an integrated theory. *IEEE Transactions on Engineering Management, 50*(2), 204–218.

Min, B.-K., & Soon Heung, C. (1991). System complexity measure in the aspect of operational difficulty. *IEEE Transactions on Nuclear Science, 38*(5), 1035–1039.

Newcomb, P. J., Bras, B., & Rosen, D. W. (1998). Implications of modularity on product design for the life cycle. *Journal of Mechanical Design, 120*(3), 483–490.

Nunnally, J. C. (1967). *Psychometric theory* (3rd ed.). New York: McGraw-Hill.

Shannon, C. E. (1948a). A mathematical theory of communication, part 1. *Bell System Technical Journal, 27*(3), 379–423.

Shannon, C. E. (1948b). A mathematical theory of communication, part 2. *Bell System Technical Journal, 27*(4), 623–656.

Shannon, C. E., & Weaver, W. (1998). *The mathematical theory of communication*. Champaign, IL: University of Illinois Press.

Simon, H. A. (1996). *The sciences of the artificial* (3rd ed.). Cambridge, MA: MIT Press.

Sosa, M. E., Eppinger, S. D., & Rowles, C. M. (2007). A network approach to define modularity of components in complex products. *Journal of Mechanical Design, 129*(11), 1118–1129.

Stevens, S. S. (1946). On the theory of scales of measurement. *Science, 103*(2684), 677–680.

Steward, D. V. (1981). The design structure system: A method for managing the design of complex systems. *IEEE Transactions on Engineering Management, EM, 28*(3), 71–74.

Suh, N. P. (1990). *The principles of design*. New York: Oxford University Press.

Suh, N. P. (2001). *Axiomatic design: Advances and applications*. New York: Oxford University Press.

Suh, N. P. (2005). *Complexity: Theory and applications*. New York: Oxford University Press.

Summers, J. D., & Shah, J. J. (2010). Mechanical engineering design complexity metrics: size, coupling, and solvability. *Journal of Mechanical Design, 132*(2), 021004.

Torgerson, W. (1958). *Theory and methods of scaling*. New York: Wiley.

Wiegers, K. E. (2003). *Software requirements* (2nd ed.). Redmond, WA: Microsoft Press.

Yourdon, E., & Constantine, L. L. (1979). *Structured design: Fundamentals of a discipline of computer design and systems design*. Englewood Cliffs, NJ: Prentice-Hall.

Yu, T.-L., Yassine, A. A., & Goldberg, D. E. (2007). An information theoretic method for developing modular architectures using genetic algorithms. *Research in Engineering Design, 18*(2), 91–109.

Chapter 7
Compatibility, Consistency, Interoperability

Abstract The design of systems and components during the design stage of the systems life cycle requires specific purposeful actions to ensure effective designs and viable systems. Designers are faced with a number of *design concerns* that they must embed into the design in every instance of thinking and documentation. Three of these concerns are addressed by the non-functional requirements for compatibility, consistency, and interoperability. Formal understanding of each of these non-functional requirements requires definitions, terms, and equations, as well as the ability to understand how to control their effect and measure their outcomes during system design endeavors.

7.1 Introduction to Compatibility, Consistency, and Interoperability

This chapter will address three major topics: (1) compatibility; (2) consistency; and (3) interoperability in design endeavors. The chapter begins by reviewing compatibility and the basic terminology, equations and concepts that underlie its utilization. Compatibility and its relation to standards is addressed and a measure for evaluating compatibility in systems design is proposed.

Section 7.2 discusses the concept of consistency and how it affects systems designs. Consistency is defined and reviewed with respect to design. The section completes with a proposed measure for consistency that is based upon requirements validation, functional verification, and design verification activities and provides a structural map relating the metric and the measurement attributes for consistency.

Section 7.5 in this chapter addresses interoperability by providing a definition and models of interoperability. A number of methods for evaluating interoperability are discussed and a measurement method is proposed. The section concludes with a metric and measurable characteristic for interoperability.

The chapter has a specific learning goal and associated objectives. The learning goal of this chapter is to be able to identify how the attributes of compatibility,

K.M. Adams, *Non-functional Requirements in Systems Analysis and Design*,
Topics in Safety, Risk, Reliability and Quality 28,
DOI 10.1007/978-3-319-18344-2_7

consistency, and interoperability that influence design in systems endeavors. This chapter's goal is supported by the following objectives:

- Define compatibility.
- Describe the terminology associated with compatibility standards.
- Construct a structural map that relates compatibility to a specific metric and measurable characteristic.
- Define consistency.
- Construct a structural map that relates consistency to a specific metric and measurable characteristic.
- Define interoperability.
- Describe the types of interoperability.
- Construct a structural map that relate interoperability to a specific metric and measurable characteristic.

The ability to achieve these objectives may be fulfilled by reviewing the materials in the sections that follow.

7.2 Compatibility

This section will review the basics of compatibility and how it is applied during systems endeavors. Compatibility is not a well-known non-functional requirement and as such must be clearly defined and understood.

7.2.1 Compatibility Definition

Compatibility, from a systems engineering perspective, is defined as:

> 1. The ability of two or more systems or components to perform their required functions while sharing the same hardware or software environment. 2. The ability of two or more systems or components to exchange information (IEEE and ISO/IEC 2010, p. 62).

The 2nd definition of compatibility is very close to the definition for interoperability which is:

> The ability of two or more systems or components to exchange information and to use the information that has been exchanged. (IEEE and ISO/IEC 2010, p. 186)

Interoperability will be discussed in the third section of this chapter, so the discussion of compatibility in this section will use only the 1st definition, where a system's ability to work with other systems without having to be altered to do so is the focus. By restricting the definition in this manner a direct linkage with the notion of standards, which are the primary means for ensuring compatibility in systems endeavors, is established.

7.2.2 *Standards—the Means for Ensuring Compatibility in Systems*

The utilization of compatible components is a fundamental requirement for every discrete system. The system must have compatible components that are able to work together in support of the system and its goals. Those responsible for system design efforts make every attempt to minimize the number of unique, one-of-a-kind components in a design in order to satisfy both sustainment concerns and contain costs. As a result, most systems are designed based on existing, commercially available, off-the-shelf components. The ability to utilize these components in a wide variety of designs is predicated upon compatibility standards. Compatibility standards "… define the interface requirements that allow different products, often from different producers, to use the same complementary goods and services, or to be connected in networks" (Grindley 1995, p. 9).

Compatibility standards are established through widespread acceptance of product specifications developed using one of the four methods described in Table 7.1.

The need for compatibility standards exists in many sectors of the global economy. It is safe to say that systems could not be affordably designed, constructed, and utilized without the existence of a large number of compatibility standards. There are hundreds of formal standards issuing bodies that issue and maintain formal standards that ensure compatibility. Very few of these bodies are administered by the government, and most are voluntary, written by trade,

Table 7.1 Types and methods of establishment for standards

Type	Method	Description
Defacto standards—emerge from market mediated forces	Unsponsored	"Sets of specifications that have no identified originator holding a proprietary interest, nor any subsequent sponsoring agency, but nevertheless exist in well-documented form in the public domain" (David and Greenstein 1990, p. 4)
	Sponsored	"One or more sponsoring entities holding a direct or indirect proprietary interest—suppliers or users, and private cooperative venture into which such firms may enter—creates inducements for other firms to adopt particular sets of technical specifications" (David and Greenstein 1990, p. 4)
Committee standards—emerge from political process	Voluntary	"Standards agreements arrived at within, and published by voluntary standards-writing organizations" (David and Greenstein 1990, p. 4)
	Mandated	"Promulgated by governmental agencies that have some regulatory authority" (David and Greenstein 1990, p. 4)

professional, and technical organizations that support specific industries and associated sectors in the economy. In the United States, the American National Standards Institute's (ANSI) stated mission is:

> To enhance both the global competitiveness of U.S. business and the U.S. quality of life by promoting and facilitating voluntary consensus standards and conformity assessment systems, and safeguarding their integrity.

Internationally, the International Organization for Standardization (ISO) and International Electrotechnical Commission (IEC) foster voluntary, industry-wide consensus for compatibility standards.

The impact that compatibility standards have is staggering. Imagine how the absence of sufficient compatibility standards would impact the daily lives of millions of people by envisioning these simple system components without applicable standards.

1. Electrical sockets
2. Electronic connectors
3. Screw threads
4. Battery sizes
5. Pipe sizes
6. Cable and wiring sizes
7. Wheel rims and tires
8. Spectrum frequency designations

In fact, systems practitioners and the engineering of systems are directly affected by many ANSI and ISO/IEC standards, some of which are listed in Table 7.2.

The adoption and utilization of compatibility standards has three clearly identified benefits and costs (Shapiro 2001).

Table 7.2 ANSI and ISO/IEC systems standards

Standard	Description
ANSI/EIA Standard 632: processes for engineering a system	Defines "what to do" with respect to the processes for engineering a system
ANSI/EIA-731, systems engineering capability	Provides a capability model and assessment method as a basis and means for determining "how well" the processes in ANSI/EIA-632 are defined and implemented"
IEEE and ISO/IEC Standard 15288: systems and software engineering—system life cycle processes.	Establishes a common process framework for describing the life cycle of man-made systems. It defines a set of processes and associated terminology for the full life cycle, including conception, development, production, utilization, support and retirement
IEEE and ISO/IEC Standard 24765: systems and software engineering—vocabulary	Provides a common vocabulary applicable to all systems and software engineering work

1. Buyers view product compatibility as a benefit because they feel they are protected from *stranding*, because the product has been designed and produced to work according to an accepted and utilized standard, protecting the buyer from the limitations associated with a one-of-a-kind product.
2. Systems designers view compatibility from the perspective that standards place limits on their design decisions. Because design decisions are constrained there are potential financial costs associated with static losses due to limited variety and dynamic losses due to limited innovation.
3. Management may view compatibility standards as both a means for muting competition during early development and extending product life by ensuring compatibility with other products over the life of the product.

In summary "Standards are an inevitable outgrowth of *systems*, whereby *complementary products work in concert to meet users' needs*" (Shapiro 2001, p. 82).

7.2.3 Compatibility in Systems Design Efforts

In the traditional systems design process invoked by *IEEE Standard 1220—Systems engineering—Application and management of the systems engineering process* (IEEE 2005) compatibility is touched upon in two process areas.

- As an element of the design verification process in section 6.6.2

  Verification evaluation where (1) methods to assure compatibility with other life cycle support functions consistent with the system design, and (2) the design element solutions satisfy the validated requirements baseline. The verification results are evaluated to ensure that the behavior exhibited by the design element solutions was anticipated and satisfies requirements.

- As an element of the control process in section 6.8.1.1.

  Interface management where interface compatibility assessments are conducted.

The most important concept to take away from each of these process activities in that the proposed design should be verified for compatibility with the prescribed design requirements and constraints.

7.2.4 Evaluating Compatibility in Design

Design Compatibility Analysis (DCA) is the process that focuses on ensuring that the proposed design is compatible with the specifications in the original design requirements. While this may seem trivial, many proposed designs stray far from the original requirements and associated specifications. The underlying goal for DCA is:

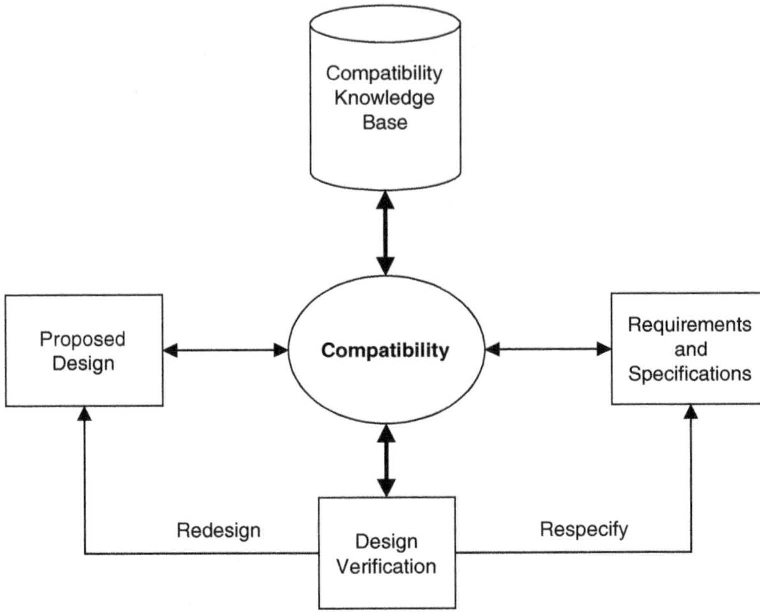

Fig. 7.1 Concept for design compatibility analysis [based on figure in (Ishii et al. 1988)]

> One of the important tasks in engineering design is to ensure the compatibility of the elements of the proposed design with each other and with the design specifications (requirements and constraints). Major design decisions, such as selection of components, determination of the system type, and sizing of components, must be made with the compatibility issue in mind. Some design features make a good match while others may be totally incompatible for the required design specifications (Ishii et al. 1988, p. 55).

The team that conducts the DCA should follow the concept depicted in Fig. 7.1 where the proposed design is compared with the requirements and specifications while using the compatibility knowledge base as a guide. If the proposed design is deemed to be non-compatible, then either a redesign or re-specification is required.

7.2.5 A Method for Measuring Compatibility in Design

The evaluation of compatibility may be accomplished using the theory of fuzzy measures (Ishii and Sugeno 1985; Zadeh 1965). In this case the fuzzy measure will be used to quantify the compatibility of the proposed design against the requirements and specifications, as depicted in Fig. 7.1. The fuzzy measure will provide a degree of confidence or a match index (MI), to evaluate the utility of the proposed design. The match index is described as follows:

> The match index is a normalized scale between 0 and 1: An MI of 0 indicates an absolutely incompatible design, an MI of 1 is a perfectly balanced design, and an MI of 0.5 indicates that no compatibility information was available. For an acceptable design, MI has to be greater than 0.5. While users may encounter MI = 0.5 if the compatibility knowledge is scarce, a more mature knowledge-base should result in some positive or negative comments about the design (Ishii et al. 1988, p. 57).

The match index is mathematically represented in Eq. 7.1.

Match Index

$$MI = \sum_{K} utility(s) * M(s), \quad s \in K \tag{7.1}$$

where:

utility(*s*) weight of the evaluation of a design element where Σ *utility*(*s*) = 1.0.
M(*s*) compatibility of a design element *s*.
K entire set of design elements.

While a full discussion and derivation of the match-index (MI) is beyond the scope of this chapter, the reader is encourage to review the literature on the use of the MI as a measure of design compatibility (Ishii 1991; Ishii et al. 1988).

7.2.6 Measuring Compatibility

At the end of each Chap. 3 the importance of being able to measure each non-functional attribute was cited as being essential in each and every system design endeavor. A structural mapping that relates compatibility to a specific metric and measurement entity are required. The four-level construct for compatibility is presented in Table 7.3.

7.3 Consistency

In this section the basics of consistency and how it is applied during systems endeavors will be discussed. Consistency is another non-functional requirements that is not well-known and as such must be clearly defined and understood.

Table 7.3 Four-level structural map for measuring compatibility

Level	Role
Concern	Systems design
Attribute	Compatibility
Metric	Design compatibility
Measurable characteristic	Design match index (*MI*)

7.3.1 Consistency Definition

Consistency, from a systems engineering perspective, is defined as:

> 1. The degree of uniformity, standardization, and freedom from contradiction among the documents or parts of a system or component 2. Software attributes that provide uniform design and implementation techniques and notations (IEEE and ISO/IEC 2010, p. 73).

Consistency may also be understood from its simplified philosophical definition—"A semantic notion: two or more statements are called consistent if they are simultaneously true under some interpretation" (Audi 1999, p. 177). In the engineering sense, where the consistency of design artifacts (i.e., a specification) are being compared and contrasted, consistency has two forms, defined in Table 7.4.

7.3.2 Consistency in Systems Design Efforts

The importance of consistency in a large system design effort cannot be overemphasized. Because elements of the design are interrelated and cascade hierarchically from stakeholder requirements to each and every lower level design artifact, the necessity for consistency arises. Consistent designs ensure that all elements of the completed design are properly represented by ensuring they do not conflict with each other or with higher level specifications or entities. A consistently designed system, when assembled and implemented, can provide predictable behaviors and improved usability for its users. In fact, recent research revealed that consistently applied language in a website had a measurable effect on users' satisfaction and ability to use the site (Ozok and Salvendy 2000).

A predictable system is one where the user has some sort of instinctive notion of what will happen as a result of their actions. A key design element that ensures that a system is predictable is consistency. In his seminal book *Designing the User Interface: Strategies for Effective Human-computer Interaction*, Ben Schneiderman (1997) lists Eight Golden Rules of Dialogue Design. Rule number one is *strive for consistency*. Schneiderman states that the following actions ensure consistency when building human-machine interfaces.

- consistent sequences of actions should be required in similar situations;
- identical terminology should be used in prompts, menus, and help screens; and
- consistent commands should be employed throughout.

Table 7.4 Forms of consistency

Form of consistency	Definition
Internal consistency	"Items within the specification do not conflict with each other" (Boehm 1984, pp. 77–78)
External consistency	"Items in the specification do not conflict with external specifications or entities" (Boehm 1984, p. 78)

Delivering a predictable systems is achieved through the use of common design principles. The design principles are statements of policy which help systems practitioners to make informed decisions during the design process. The design principles contain high level guidance which requires comprehension before application in real-world design scenarios.

Despite the obvious need for consistent designs, neither the term consistency nor the word predictable are directly addressed in *IEEE Standard 1220—Systems engineering—Application and management of the systems engineering process* (IEEE 2005). However, the reason that consistency is a feature in the purposeful design of systems is because, from an engineering design perspective consistency has one major purpose—it ensures that each of the models that make up the overall engineering design yield design artifacts that are simultaneously true and represent the same design model. This is traditionally accomplished as part of the systems validation and verification processes as described in the following.

- As an element of the requirements validation process in Sect. 6.2.4.
  Variances and conflicts in the system's requirements are identified and resolved by iterating through requirements analysis to correct the requirements baseline.
- As an element of the functional verification process in Sect. 6.4.3.
  Variances and conflicts in the system's functions, functional architecture, measures of performance, and constraints are identified and resolved by iterating back through functional requirements and requirements analysis to correct the verified functional architecture.
- As an element of the design verification process in section 6.6.3.
  Variances and conflicts in the system's design are identified and resolved by iterating back through synthesis and functional analysis to correct the design elements.

7.3.3 Methods for Evaluating Consistency in Design

"Assessing consistency is really a process for ensuring that the different viewpoint models … form projections from the same overall design model" (Budgen 2003, p. 384). There is a dearth of information in the literature on methods and techniques for *how* to assess consistency. Completion of the requirements validation (6.2.4), functional verification (6.4.3), and design verification (6.6.3) tasks in *IEEE Standard 1220* (IEEE 2005) ensure that consistency is addressed as a high level task. Boehm (1984) recommended using the following:

- *Manual cross-referencing*: Cross referencing involves both reading and the construction of cross-reference tables and diagrams to clarify interactions among design entities. For many large systems this can be cumbersome, leading to the suggested use of automated cross-referencing tools.

- *Automated Cross-referencing:* Automated cross-referencing involves the use of a machine to analyze the language contained in the design artifacts.
- *Interviews*: Discussing a design artifact with its designer is a logical way to identify or clear up potential problems. Interviews are particularly good for resolving understandings and addressing high-risk issues.
- *Checklists*: Specialized lists, based on the experience of the design verification and validation team, can be used to ensure that significant issues are addressed during the team's review. Checklists are utilized by the design review team as a guide and do not represent additional systems requirements.
- *Simple Scenarios*: Scenarios describe how the system will work once it is in operation and are very good for clarifying misunderstandings or mismatches between the specification and the design artifacts.
- *Detailed Scenarios:* Scenarios with more elaborate and expansive systems operational descriptions are often more effective than simple scenarios. However, the cost associated with detailed scenarios is often prohibitive during design reviews and often recommended for inclusion as part of the post-production test and evaluation stage.
- *Prototypes:* Some design artifacts are so abstract that their feasibility cannot be commented on without the development of a working prototype. As with detailed scenarios, the cost associated with detailed scenarios is highly prohibitive during design reviews. Despite this, prototypes are necessary to eliminate design ambiguities and inconsistencies in the design and are required prior to the production stage.

7.3.4 A Method for Measuring Consistency in Design

The previous sections defined consistency, indicated where it is addressed during system design endeavors, and discussed some methods for evaluating consistency in a design. The measurement goal is to ensure that consistency is treated as a purposeful element of the design process. In order to ensure that the system design process has invoked practices within the design effort that ensure consistency, the design effort should include relevant measures. As with traceability, the criteria for consistency will be subjective, qualitative measures that will need to answer questions that address both the presence (yes or no) and quality of the effort (how well) at ensuring consistency in each of the three design processes we have identified. In this case consistency methods will need to relate to a specific measurable attribute that can be utilized as a measure. Once again, measures are important because they are the linkage between observable, real-world, empirical facts and the construct we create as an evaluation point.

7.3.4.1 Scale for Design Consistency

As discussed during the development of the scales for traceability in the previous chapter, the selection of a measurement scale is an important element in the development of an adequate measure for design consistency. Because the three design processes (i.e., requirements validation, functional verification, and design verification) used to identify evaluation points have no natural origin or empirically defined distance, the ordinal scale is selected as an appropriate scale for measuring the consistency attributes. The well-known Likert scale is proposed for use in evaluating design consistency. In order to improve its reliability a five point Likert scale will be invoked (Lissitz and Green 1975).

Before moving on to describing the measure for design consistency two important points must once again be made with respect to scale development. Scales are characterized as either a *proposed* scale or a scale. "A proposed scale is one that some investigator(s) put forward as having the requisite properties, and if it is indeed shown to have them, then it is recognized as a scale" (Cliff 1993, p. 65). As stated earlier, the use of the word scale is referring to *proposed scales*. This may seem to be an insignificant point, but until the scale has been accepted and successfully utilized it remains proposed.

7.3.4.2 Proposed Measurement Scale for Design Consistency

Armed with a construct, measurement attributes and an appropriate scale type, the design consistency measure may be constructed. In order to evaluate design consistency, questions that address both the presence (yes or no) and quality of the effort (how well) must be answered in order to provide consistent systems design by evaluating design processes for requirements validation, functional verification, and design verification tasks. These measurement constructs are presented in Table 7.5.

Table 7.5 Measurement constructs for consistency

Life cycle stage or process	IEEE Std 1220 section	Consistency concern for measurement
Requirements validation process	6.2.4	*Variances and conflicts in the system's requirements are identified and resolved by iterating through requirements analysis to correct the requirements baseline*
Functional verification process	6.4.3	*Variances and conflicts in the system's functions, functional architecture, measures of performance, and constraints are identified and resolved by iterating back through functional requirements and requirements analysis to correct the verified functional architecture*
Design verification process	6.6.3	*Variances and conflicts in the system's design are identified and resolved by iterating back through synthesis and functional analysis to correct the design elements*

Table 7.6 Measurement questions for design consistency

Measurement construct	Consistency concern for measurement
C_{rv}	*Are variances and conflicts in the system's requirements identified and resolved by iterating through requirements analysis to correct the requirements baseline?*
C_{fv}	*Are variances and conflicts in the system's functions, functional architecture, measures of performance, and constraints are identified and resolved by iterating back through functional requirements and requirements analysis to correct the verified functional architecture?*
C_{dv}	*Are variances and conflicts in the system's design are identified and resolved by iterating back through synthesis and functional analysis to correct the design elements?*

Table 7.7 Consistency measurement question Likert scale

Measure	Descriptor	Measurement criteria
0.0	None	No objective quality evidence is present
0.5	Limited	Limited objective quality evidence is present
1.0	Nominal	Nominal objective quality evidence is present
1.5	Wide	Wide objective quality evidence is present
2.0	Extensive	Extensive objective quality evidence is present

In order to evaluate the design's ability to conform to the notion of consistency, a specific question should be developed which will evaluate the appropriate areas in each design process. The measurement constructs and questions associated with each of the consistency measurements concerns are presented in Table 7.6.

The answer to each question in Table 7.6 will be scored using the 5-point Likert measures in Table 7.7.

The overall measure for system consistency is a sum of the scores from the nine traceability metrics as shown in Eqs. 7.2 and 7.3.

Generalized Equation for System Consistency

$$C_{sys} = \sum_{i=1}^{n} C_i \tag{7.2}$$

A generalized measure for system consistency is shown in Eq. 7.2.

Expanded Equation for Systems Consistency

$$C_{sys} = C_{rv} + C_{fv} + C_{dv} \tag{7.3}$$

The summation of the three constructs in Eq. 7.3 will be the measure the degree of consistency in a system design endeavor.

7.3.5 Measuring Consistency

At the end of Chap. 3 the importance of being able to measure each non-functional attribute during systems design endeavors was stressed. A structural mapping that relates consistency to a specific metric and measurement entity are required. The four-level construct for consistency is presented in Table 7.8.

7.4 Interoperability

In this section the basics of interoperability and how it is applied during systems endeavors will be addressed. Interoperability is a common term, but one without a generally agreed upon definition. As such, it must be clearly defined and understood.

7.4.1 Interoperability Definition

Interoperability, from a systems engineering perspective, is defined as.

> The ability of two or more systems or components to exchange information and to use the information that has been exchanged (IEEE and ISO/IEC 2010, p. 186).

Interoperability, as a non-functional requirement, has additional definitions, shown in Table 7.9 that may provide improved understanding of the term.

The definition for interoperability can be further refined by reviewing the characteristics for seven types of interoperability reported in the extant literature.

1. *Technical interoperability:* concerns the ability to exchange services or information directly and satisfactorily between systems and their users (Kinder 2003).
2. *Syntactic interoperability*: addresses the includes the ability to deal with formatting and data exchange as supported by standards (Sheth 1999).
3. *Semantic interoperability*: ensures that data embedded within services or information are interpreted by senders and receivers as representing the same

Table 7.8 Four-level structural map for measuring consistency

Level	Role
Concern	Systems design
Attribute	Consistency
Metric	Design consistency
Measurable characteristic	Requirements validation consistency (C_{rv}), Functional verification consistency (C_{fv})
	Design verification consistency (C_{dv})

Table 7.9 Additional definitions for interoperability

Definition	Source
"The ability to exchange services and data with one another"	Heiler (1995, p. 271)
"Interoperability is the ability of systems, units, or forces to provide services to and accept services from other systems, units, or forces and to use the services exchanged to enable them to operate effectively together"	Kasunic and Anderson (2004, p. vii)
"The ability by which the elements of a system can exchange and understand the required information with each other"	Rezaei et al. (2014, p. 22)

concepts, relations, or entities, in a suitable abstraction of the real-world (Vetere and Lenzerini 2005).

4. *Organizational interoperability:* addresses the definition of authority and responsibility for the exchange of services or information (Chiu et al. 2004).
5. *Programmatic interoperability:* concerns the relationships between the organizational entities responsible for managing the systems that require interoperable services or data (DiMario 2006).
6. *Constructive interoperability*: addresses the design elements that govern development of the architecture, standards, and engineering for the systems that require interoperable services or data (DiMario 2006).
7. *Operational interoperability*: addresses the technical ability of the systems to interact with one another and the environment (DiMario 2006).

From the seven type descriptions it is clear that some of these types are both similar and have overlapping elements. In the next section interoperability several models for interoperability will be reviewed.

7.4.2 Models for Interoperability

Interoperability in systems may also be addressed by viewing how interoperability concerns have shifted over time. Three distinct periods, or interoperability generations have occurred (Sheth 1999) and may be represented by some of the characteristics in Table 7.10.

Table 7.10 Characteristics of interoperability generations

Characteristic	Generation I	Generation II	Generation III
Major concern	Data	Information	Knowledge
Interoperability emphasis	Structural	Syntax	Semantic
Interoperability scope	Applications	Small numbers of system	Enterprise-wide
Interoperability techniques	Data relationships	Single ontology	Multiple ontologies
Types of data	Files	Structured databases	Multi-media

Table 7.11 LCIM levels and capability concepts

Capability	Interoperability level and title	Description
Composability	6—Conceptual	The assumptions and constraints of the meaningful abstraction of reality—are aligned between participating systems
	5—Dynamic	Participating systems are able to comprehend the state changes that occur in the assumptions and constraints that each other is making over time
Interoperability	4—Pragmatic	The context in which the information is exchanged is unambiguously defined
	3—Semantic	The meaning of the data is shared; the content of the information exchange requests are unambiguously defined
	2—Syntactic	A common protocol to structure the data is used; the format of the information exchange is unambiguously defined
Integratability	1—Technical	A communication protocol exists for exchanging data between participating systems
	0—None	Stand-alone systems with no interoperability

An extensive model of interoperability has been developed to support modeling and simulation efforts that shows promise for supporting higher levels of interoperability. The Levels of Conceptual Interoperability Model (LCIM) introduces technical, syntactic, semantic, pragmatic, dynamic, and conceptual layers of interoperation and shows how they are related to the increasing capability represented by integratability, interoperability, and composability (Tolk et al. 2007). Table 7.11 describes the levels and increasing capabilities in the LCIM.

A holistic perspective for interoperability in systems may be represented as the locus of the three generations and the LCIM. The holistic perspective is termed *systemic interoperability* and it moves beyond the traditional, mechanistic or *hard* interoperability represented by structure, syntax, and semantics. The new perspective for systemic interoperability "… is a holistic view of interoperability and requires compatibility in worldview and conceptual, contextual, and cultural interoperability…" (Adams and Meyers 2011, p. 172) as depicted in Fig. 7.2.

7.4.3 Interoperability in Systems Design Efforts

Interoperability is not directly addressed in *IEEE Standard 1220—Systems engineering—Application and management of the systems engineering process* (IEEE 2005). However, it is indirectly addressed by the sustainment concern *operability*. If interoperability is included where operability is mentioned, then interoperability is addressed in two areas.

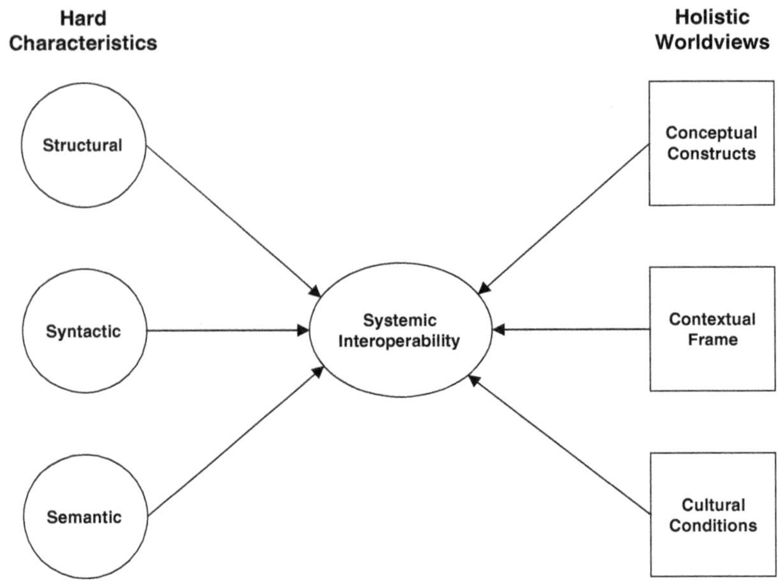

Fig. 7.2 Movement from hard to systemic interoperability

- As an element of the modeling and prototyping task in Sect. 4.5.

  Suitable models, simulations, or prototypes should be developed and utilized to evaluate interoperability.

- As an element of the define measures of effectiveness task in Sect. 6.1.5.

  Define system effectiveness measures that reflect overall stakeholder expectations and satisfaction which may include interoperability.

7.4.4 Methods for Evaluating Interoperability

Evaluation of interoperability and development of an appropriate metric and supporting measurement criteria is a non-trivial task. Early efforts within the United States Department of Defense summarized the situation as:

> Interoperability is a broad and complex subject. Developing and applying precise measurements in an area as multidimensional and complex as interoperability is difficult. However, measuring, assessing, and reporting interoperability in a visible way is essential to setting the right priorities (Kasunic and Anderson 2004, p. vii).

During the period between 1980 and 2014 sixteen separate models for evaluating interoperability in systems have appeared in the general literature. However only eight of these are either major enterprise reports or have been published in peer-reviewed scholarly journals. Table 7.12 presents the eight formal models.

Table 7.12 Peer-reviewed models for evaluating systems interoperability

Model and brief description	Reference
Spectrum of Interoperability Model (SoIM)—A simple two element model that evaluates interoperability in terms of technical possibility and management/control possibility	LaVean (1980)
Quantification of Interoperability Methodology (QoIM)—A model that associates interoperability with measures of effectiveness	Mensh et al. (1989)
Levels of Information Systems Interoperability (LISI)—A maturity model that considers five levels of increasing sophistication with respect to exchanging and sharing information and services	DoD (1998)
Stop Light Model—A simple, readiness reporting style method of measuring interoperability	Hamilton et al. (2002)
Net Centric Warfare Maturity Model (NCW)—A five-level maturity model that models the maturity of situational awareness and command and control in the context of five interoperability levels	Alberts and Hayes (2003)
Levels of Conceptual Interoperability Model (LCIM)—A seven level model that evaluates interoperability in terms of integration, interoperability and the ultimate goal of composability	Tolk et al. (2007)
i-Score—A complex model that evaluates systems according to their interoperability-related features in the context of an operational process	Ford et al. (2009)
Ultra large scale systems interoperability maturity model—A five-level maturity model for evaluating ultra large scale systems that utilizes technical, syntactic, semantic, and organizational interoperability domains	Rezaei et al. (2014)

While an in-depth review of each model is beyond this chapter, readers are encouraged to consult the references in Table 7.12 for a more detailed description of the development of each interoperability evaluation model. The next section will describe one of these measures as an appropriate technique for measuring and evaluating interoperability in systems.

7.4.5 i-Score Model for Evaluating System Interoperability

The *i-Score* Model for evaluating system interoperability compares the similarity of system interoperability characteristics and is based upon the following assumption:

> If a pair of systems is instantiated only with system interoperability characters, then the measure of their similarity is also a measure of their interoperability (Ford 2008, p. 52).

The interoperability example utilized to describe this technique includes the macro-system S which is made up of the set of systems s_i, which includes the macro-characteristics of interoperability X, which are made up of the individual system interoperability characteristics x_i. For two systems s_1 and s_2, we can have the interoperability relationships $X(s)$ as shown in Fig. 7.3.

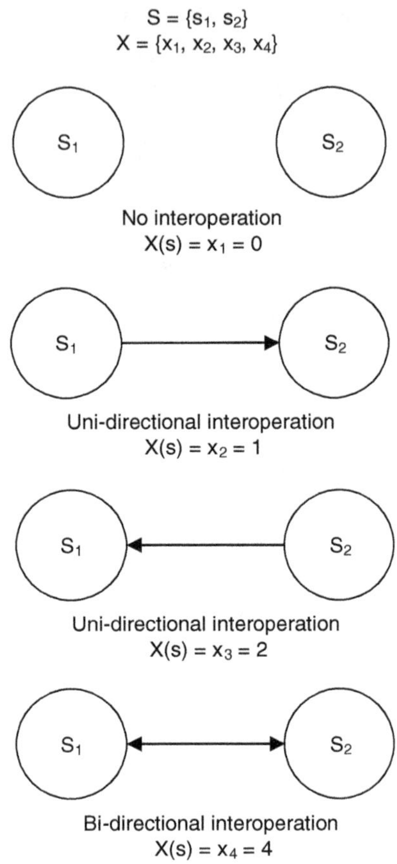

Fig. 7.3 Directional interoperability types

The symbols used in Fig. 7.3 are defined as follows:

- S = A set of n systems s_i, where i = 1 to n and $s_i \in S$
- s_i = An individual system s_i, where i = 1 to n
- X = A set of n system characteristics X_i, where i = 1 to n and $x_i \in X$
- x_i = A system characteristic x_i, where i = 1 to n.
- C = A set of n character states c_i, where i = *1* to n and $c_i \in C$
- c_i = The character states c_i, where i = *1* to n

The character states are the actual states that the measurable interoperability characteristics may have. For the system depicted in Fig. 7.3 there are four possible character states: (1) no interoperation; (2) send left-to-right; (3) send right-to-left; and (4) send both ways.

These system characteristics (x_i) represent measurable attributes which may be used to describe an important interoperability feature of the system. The individual system characteristics combine to create an overall representation of the characteristics of the system. The systems interoperability characteristics include four generalized types shown in Table 7.13 (Ford 2008).

Table 7.13 Types and examples of interoperability characteristics

Type of characteristic	Examples
Morphological	Size, shape, color, structure, number of components, etc
Physiological	Function, behavior, etc
Ecological	Context, environment, resource consumption, etc
Distributional	Geographic location, domain, etc

Most importantly, the interoperability characteristic must have a natural measure that ensures the characteristic is unique, reliable (e.g., able to be repeatedly measured), unambiguous, and representational.

7.4.5.1 *i*-Score System Interoperability Evaluation Equation

Once the set *(S)* of systems (s_i), their interoperability characters (x_i), and the possible states of those characters (c_i) are identified, modeling of the interoperability relationships for the larger system *(S)* may commence. The individual systems (s_i) are modeled, or instantiated, as a sequence that is representative of the states of each system's interoperability characteristics. The lower-case Greek character sigma (σ) is used to denote the instantiation of the individual system (s_i), as depicted in Eq. 7.4.

Instantiation of s_i

$$\sigma(\mathrm{s_i}) = \{x_1(s_1), x_2(s_1), \ldots x_n(s_1)\} \tag{7.4}$$

In order to support meaningful system comparisons the larger system (*S*) is modeled by aligning the instantiations (σ_i) of all of the member systems $(s_i \in S)$. The upper-case Greek characters sigma (Σ) is used to denote the instantiation alignment as depicted in Eq. 7.5.

Instantiation Alignment of *S*

$$\sum_{i=1}^{n} X_i(S_i) = \{\sigma_1, \sigma_2, \ldots \sigma_n\} \tag{7.5}$$

The result of Eq. 7.5 is a matrix of the following form.

$$\begin{matrix} x_1(s_1), x_2(s_1), \ldots x_n(s_1) \\ x_1(s_2), x_2(s_2), \ldots x_n(s_2) \\ x_1(s_n), x_2(s_n), \ldots x_n(s_n) \end{matrix}$$

An interoperability function (*I*), shown in Eq. 7.6, has been proposed by Ford (2008) that uses the modified Minkowski similarity function to derive a weighted, normalized measure of the similarity of two system instantiations (σ′) and (σ″).

Interoperability Function

$$I = \left[\frac{\sum_{i=1}^{n} \sigma'(i) + \sum_{i=1}^{n} \sigma''(i)}{2nc_{\max}}\right]\left[1 - \left(\frac{1}{\sqrt[r]{n}}\right)\left[\sum_{i=1}^{n} b_i \left[\frac{\sigma'(i) - \sigma\prime\prime(i)}{c_{\max}}\right]^{r}\right]^{\frac{1}{r}}\right] \quad (7.6)$$

where:

r is the Minkowski parameter (usually set to $r = 2$).
n the number of interoperability characters in each system.
b 0, if $\sigma'(i) = 0$ or $\sigma''(i) = 0$, else $b = 1$.
c_{max} maximum value of any interoperability character.

7.4.5.2 i-Score Example of System Interoperability Evaluation

Three systems s_1, s_2, and s_3 where (s_1, s_2, $s_3 \in S$) have interoperability characteristics x_1, x_2, x_3, and x_4 where (x_1, x_2, x_3, $x_4 \in X$) and the characteristics have a maximum value of 9 where $C \in \{R \cap [0,9]\}$ with r = 2) can be represented as follows:

$$S = \{s_1, s_2, s_3\}$$
$$X = \{x_1, x_2, x_3, x_4\}$$
$$\{\sigma_1, \sigma_2, \sigma_3\} = \{x_1(s_1), x_2(s_1), x_3(s_1), x_4(s_1); x_1(s_2), x_2(s_2), x_3(s_2), x_4(s_2); x_1(s_3), x_2(s_3), x_3(s_3), x_4(s_3)\}$$

Σ is the aligned instantiation where Σ = *X(S)* and is represented by the matrix:

$$\sum = X(S) = \begin{matrix} x_{1(s1)} & x_{2(s1)} & x_{3(s1)} & x_{4(s1)} \\ x_{1(s2)} & x_{2(s2)} & x_{3(32)} & x_{4(s2)} \\ x_{1(s3)} & x_{2(s3)} & x_{3(s3)} & x_{4(s3)} \end{matrix}$$

Our example involves the system *X(S)* represented by Σ

$$\sum = \begin{vmatrix} 1 & 2 & 3 & 0 \\ 4 & 5 & 6 & 0 \\ 7 & 8 & 9 & 0 \end{vmatrix}$$

The interoperability function *I* may be calculated by inserting the values from Σ into Eq. 7.6. The resulting interoperability matrix *M* is:

$$M = \begin{vmatrix} 0 & 0.207 & 0.162 \\ 0.207 & 0 & 0.276 \\ 0.162 & 0.276 & 0 \end{vmatrix}$$

Table 7.14 Four-level structural map for measuring interoperability

Level	Role
Concern	Systems design
Attribute	Interoperability
Metric	System interoperability function, *I*
Measurable characteristic	System interoperability matrix, *M*

The system interoperability matrix *M* is used as a measure of the interoperability between systems in *S* where ($s_{1,}$ s_2 ... $s_n \in S$) which have interoperability characteristics *x* where ($x_{1,}$ $x_2 ... x_n \in X$).

7.4.6 Measuring Interoperability

Chapter 3 stressed the importance: of being able to measure each non-functional attribute. A structural mapping that relates interoperability to a specific metric and measurement entity are required. The four-level construct for interoperability is presented in Table 7.14.

7.5 Summary

In this chapter the non-functional requirements for compatibility, consistency, and interoperability have been reviewed. A formal definition for each non-functional requirement has been provided along with additional explanatory definitions, terms, and equations. The ability to effect the non-functional requirement during the design process has also been addressed. Finally, a formal metric and measurement characteristic have been proposed for evaluating each non-functional requirement attribute.

The chapter that follows will address non-functional requirement for safety as the final element in design concerns in systems endeavors.

References

Adams, K. M., & Meyers, T. J. (2011). Perspective 1 of the SoSE methodology: Framing the system under study. *International Journal of System of Systems Engineering, 2*(2/3), 163–192.

Alberts, D. S., & Hayes, R. E. (2003). *Power to the edge: Command and control in the information age*. Washington, DC: DoD Command and Control Research Program.

Audi, R. (Ed.). (1999). *Cambridge dictionary of philosophy* (2nd ed.). New York: Cambridge University Press.

Boehm, B. W. (1984). Verifying and validating software requirements and design specifications. *IEEE Software, 1*(1), 75–88.

Budgen, D. (2003). *Software design* (2nd ed.). New York: Pearson Education.

Chiu, D. K. W., Cheung, S. C., Till, S., Karlapalem, K., Li, Q., & Kafeza, E. (2004). Workflow view driven cross-organizational interoperability in a web service environment. *Information Technology and Management, 5*(3–4), 221–250.

Cliff, N. (1993). What is and isn't measurement. In G. Keren & C. Lewis (Eds.), *A handbook for data analysis in the behavioral sciences: Methodological issues* (pp. 59–93). Hillsdale, NJ: Lawrence Erlbaum Associates.

David, P. A., & Greenstein, S. (1990). The economics of compatibility standards: An introduction to recent research. *Economics of Innovation and New Technology, 1*(1–2), 3–41.

DiMario, M. J. (2006). System of systems interoperability types and characteristics in joint command and control. In Proceedings *of the 2006 IEEE/SMC International Conference on System of Systems Engineering* (pp. 236–241). Piscataway, NJ: Institute of Electrical and Electronics Engineers.

DoD. (1998). *C4ISR Architecture working group final report—levels of information system interoperability (LISI).* Washington, DC: Department of Defense.

Ford, T. C. (2008). *Interoperability measurement. Air force institute of technology.* Fairborn, OH: Wright Patterson Air Force Base.

Ford, T. C., Colombi, J. M., Jacques, D. R., & Graham, S. R. (2009). A general method of measuring interoperability and describing its impact on operational effectiveness. *The Journal of Defense Modeling and Simulation: Applications, Methodology, Technology, 6*(1), 17–32.

Grindley, P. (1995). *Standards, strategy, and policy: Cases and stories.* New York: Oxford University Press.

Hamilton, J. A., Rosen, J. D., & Summers, P. A. (2002). An interoperability roadmap for C4ISR legacy systems. *Acquisition Review Quarterly, 28*, 17–31.

Heiler, S. (1995). Semantic interoperability. *ACM Computing Surveys, 27*(2), 271–273.

IEEE. (2005). *IEEE Standard 1220: Systems engineering—application and management of the systems engineering process.* New York: Institute of Electrical and Electronics Engineers.

IEEE, & ISO/IEC. (2010). IEEE and ISO/IEC Standard 24765: Systems and software engineering —vocabulary. New York and Geneva: Institute of Electrical and Electronics Engineers and the International Organization for Standardization and the International Electrotechnical Commission.

Ishii, K. (1991). Life-cycle engineering using design compatibility analysis. In *Proceedings of the 1991 NSF Design and Manufacturing Systems Conference* (pp. 1059–1065). Dearborn, MI: Society of Manufacturing Engineers.

Ishii, K., Adler, R., & Barkan, P. (1988). Application of design compatibility analysis to simultaneous engineering. *Artificial Intelligence for Engineering Design, Analysis and Manufacturing, 2*(1), 53–65.

Ishii, K., & Sugeno, M. (1985). A model of human evaluation process using fuzzy measure. *International Journal of Man-Machine Studies, 22*(1), 19–38.

Kasunic, M., & Anderson, W. (2004). *Measuring systems interoperability: Challenges and opportunities (CMU/SEI-2004-TN-003).* Pittsburgh, PA: Carnegie Mellon University.

Kinder, T. (2003). Mrs Miller moves house: The interoperability of local public services in europe. *Journal of European Social Policy, 13*(2), 141–157.

LaVean, G. E. (1980). Interoperability in defense communications. *IEEE Transactions on Communications, 28*(9), 1445–1455.

Lissitz, R. W., & Green, S. B. (1975). Effect of the number of scale points on reliability: A Monte Carlo approach. *Journal of Applied Psychology, 60*(1), 10–13.

Mensh, D., Kite, R., & Darby, P. (1989). A methodology for quantifying interoperability. *Naval Engineers Journal, 101*(3), 251–259.

Ozok, A. A., & Salvendy, G. (2000). Measuring consistency of web page design and its effects on performance and satisfaction. *Ergonomics, 43*(4), 443–460.

Rezaei, R., Chiew, T. K., & Lee, S. P. (2014). An interoperability model for ultra large scale systems. *Advances in Engineering Software, 67*, 22–46.

Shapiro, C. (2001). Setting compatibility standards: Cooperation or collusion? In R. C. Dreyfuss, D. L. Zimmerman, & H. First (Eds.), *Expanding the boundaries of intellectual property: Innovation policy for the knowledge society* (pp. 81–101). New York: Oxford University Press.

Sheth, A. P. (1999). Changing focus on interoperability in information systems: From system, syntax, structure to semantics. In M. Goodchild, M. Egenhofer, R. Fegeas, & C. Kottman (Eds.), *Interoperating geographic information systems* (pp. 5–29). New York: Springer.

Shneiderman, B. (1997). *Designing the user interface: Strategies for effective human-computer interaction* (3rd ed.). Boston: Addison-Wesley.

Tolk, A., Diallo, S. Y., & Turnitsa, C. D. (2007). Applying the levels of conceptual interoperability model in support of integratability, interoperability, and composability for system-of-systems engineering. *Journal of Systemics, Cybernetics and Informatics, 5*(5), 65–74.

Vetere, G., & Lenzerini, M. (2005). Models for semantic interoperability in service-oriented architectures. *IBM Systems Journal, 44*(4), 887–903.

Zadeh, L. A. (1965). Fuzzy sets. *Information and Control, 8*(3), 338–353.

Chapter 8
System Safety

Abstract The design of systems and components during the design stage of the systems life cycle requires specific purposeful actions to ensure effective designs and viable systems. Designers are faced with a number of *design concerns* that they must embed into the design in every instance of thinking and documentation. Safety is one of these concerns and is addressed by the non-functional requirement for safety which is composed of seven attributes. The development of the seven safety attributes were developed using Leveson's Systems-Theoretic Accident Model and Processes (STAMP). Because STAMP is a system-theoretic approach appropriate for evaluating safety in complex, systems-age engineering systems, the safety attributes provide the ability to understand how to control safety and measure its outcomes during system design endeavors.

8.1 Introduction to Safety

This chapter will address system safety and how it is incorporated into system design endeavors. Machine age systems safety is contrasted with systems-age concerns. The need for safety expressed in the *IEEE Standard for the Application and Management of the Systems Engineering Process* (IEEE 2005) is used to develop a metric for evaluating safety in systems designs. The chapter completes by relating the proposed measure for evaluating systems safety as a metric and includes a structural map for systems safety.

The chapter has a specific learning goal and associated objectives. The learning goal of this chapter is to be able to identify how the attributes of safety are ensured through purposeful design efforts during systems endeavors. This chapter's goal is supported by the following objectives:

- Define safety in terms of emergence.
- Describe the relationship between systems safety and hazards.

K.M. Adams, *Non-functional Requirements in Systems Analysis and Design*,
Topics in Safety, Risk, Reliability and Quality 28,
DOI 10.1007/978-3-319-18344-2_8

Table 8.1 Additional definitions for safety

Definition	Source
"Freedom from accidents (loss events)"	Leveson (2011, p. 467)
"An emergent property that arises when the system components interact within an environment. Emergent properties like safety are controlled or enforced by a set of constraints (control laws) related to the behavior of the system components"	Leveson (2011, p. 67)
"Methods and techniques of avoiding accident or disease"	Parker (1994, p. 431)

- Describe the difference between machine-age and systems-age safety concerns.
- Construct a structural map that relate systems safety to a specific metric and measurable characteristic.

The ability to achieve these objectives may be fulfilled by reviewing the materials in the sections that follow.

8.2 Safety Definition

Safety is a widely used term, but one which we will need to define clearly if we are to apply it as a valid design concern during systems endeavors. Safety, from a systems engineering perspective, is defined as:

> The expectation that a system does not, under defined conditions, lead to a state in which human life, health, property, or the environment is endangered. (IEEE and ISO/IEC 2010, p. 315)

Safety has additional definitions, shown in Table 8.1 that may provide improved understanding of the term when used as a non-functional requirement for a system. The definitions for safety can be further improved by reviewing how safety has been addressed in the literature on systems.

8.3 Safety in Systems

Safety is like motherhood and apple pie, who does not want it? However, undertaking systems endeavors and ensuring that the associated system is safe is not an abstract idea but a concrete requirement, most often satisfied by the inclusion of a non-functional requirement for safety.

Most of the traditional literature on systems safety has focused on assumptions that worked well for simple systems (i.e., the machine age), but no longer work for complex systems (i.e., the systems age). In her seminal work *Engineering a Safer World: Systems Thinking Applied to Safety,* Leveson (2011) of the Massachusetts Institute of Technology, discusses the need to move away from the assumptions of

Table 8.2 Machine age and systems age safety model assumptions (Leveson 2011, p. 57)

	Machine age assumption	Improved systems age assumption
1	"Safety is increased by increasing system or component reliability. If components or systems do not fail, then accidents will not occur"	"High reliability is neither necessary nor sufficient for safety"
2	"Accidents are caused by chains of directly related events. We can understand accidents and assess risk by looking at the chain of events leading to the loss"	"Accidents are complex processes involving the entire socio-technical system. Traditional event-chain models cannot describe this process adequately"
3	"Probabilistic risk analysis based on event chains is the best way to assess and communicate safety and risk information"	"Risk and safety may be best understood and communicated in ways other than probabilistic risk analysis"
4	"Most accidents are caused by operator error. Rewarding safe behavior and punishing unsafe behavior will eliminate or reduce accidents significantly"	"Operator behavior is a product of the environment in which it occurs. To reduce operator 'error' we must change the environment in which the operator works"
5	"Highly reliable software is safe"	"Highly reliable software is not necessarily safe. Increasing software reliability or reducing implementation errors will little impact on safety"
6	"Major accidents occur from the chance simultaneous occurrence of random events"	"System will tend to migrate toward states of higher risk. Such migration is predictable and can be prevented by appropriate system design or detected during operations using leading indicators of increasing risk"
7	"Assigning blame is necessary to learn from and prevent accidents or incidents"	"Blame is the enemy of safety. Focus should be on understanding how the system behavior as a whole contributed to the loss and not on who or what to blame for it"

the machine age and to a new series of improved assumptions that can be successfully applied as a new model of systems safety appropriate for the systems age.[1] Table 8.2 contrasts the machine age assumptions with those required in the new systems age.

The foundation for systems engineering is systems theory and its series of supporting axioms and principles (Adams et al. 2014). Leveson's new model for system safety makes use of system theory's centrality axiom and its pair of supporting principles—hierarchy and emergence and communications and control. Specifically, safety can be viewed as a control problem:

[1]The reader is encouraged to read Chap. 2—Questioning the Foundations of Traditional Safety Engineering in Leveson (2011). *Engineering a Safer World: Systems Thinking Applied to Safety*. Cambridge, MA: MIT Press for a thorough discussion of the rationale associated with each of the seven assumptions.

> Emergent properties like safety are controlled or enforced by a set of constraints (control laws) related to the behavior of the system components. (Leveson 2011, p. 67)

Accidents in systems occur when component failures, environmental disturbances, and malfunctions among system components are not adequately controlled. In systems-age, complex systems (i.e., tightly coupled with interactive complexity) accidents are a result of inadequate control and are labeled normal accidents.

> The odd term normal accident is meant to signal that, given the system characteristics, multiple and unexplained interactions of failures are inevitable. This is an expression of an integral characteristic of the system, not a statement of frequency. (Perrow 1999, p. 5)

The interactive complexity and tight coupling in modern complex systems requires the new systems-age model of systems safety to approach safety by incorporating both the social and technical elements of the system. The socio-technical system's context will dictate the non-functional requirements for safety that are invoked during the systems design process.

8.4 Safety in System Design Efforts

In the system design process safety requirements are derived from accident hazards.

> A safety requirement is a constraint derived from identified hazards. (Penzenstadler et al. 2014, p. 42)

The definition of hazards is based upon the system design. The major elements of the design that give insight into potential hazards include the (1) system components, (2) component interconnections, (3) human interactions with the system, (4) connections to the environment, and (5) potential environmental disturbances. The formal design process should address each of these hazards and the constraint that may prevent their occurrence.

In the traditional systems design process invoked by *IEEE Standard 1220—Systems engineering—Application and management of the systems engineering process* (IEEE 2005) safety is touched upon in four process areas.

- As an element of the requirements analysis process in the following sections:
  6.1.1—Stakeholder expectations are balanced with an analysis of the effects on the overall system design and safety.
  6.1.4—Measures of effectiveness are defined reflect overall stakeholder expectations and satisfaction that include safety.
  6.1.9.5—The design team accounts for the system design features that create significant hazards.
- As an element of the functional analysis process in the following sections:
  6.3.2.5 (1)—The design team analyzes and prioritizes potential functional failure modes to define failure effects and identify the need for fault detection and recovery functions.

6.3.2.5 (2)—Functional reliability models are established to support the analysis of system effectiveness for each operational scenario.
6.3.2.5 (3)—Failures, which represent significant safety hazards, are modeled to completely understand system impacts.
6.3.2.6 (1)—The design team analyzes subfunctions to identify operational hazards.
6.3.2.6 (2)—Additional functional requirements are derived and defined for monitoring dangerous operational conditions, or notifying or warning operators of impending hazards.
- As an element of the synthesis process in the following sections:
  6.5.3—The design team analyzes all design to identify potential hazards to the system, humans involved in the system and supporting the system life cycle processes, or the environment.
- As an element of the systems analysis process in the following section:
  6.7.6.3—The design team identifies safety impacts associated with system implementation. Safety laws and regulations should be identified, and the design team should ensure that these are complied with by each solution alternative.

The most important concept to take away from each of these traditional activities in that the formal design process should be focused upon potential hazards and the constraint that may prevent their occurrence.

It is important to note that a number formal standards for systems safety exist (Alberico et al. 1999; DoD 2000; IEEE 1994; NASA 2011). However, most of these standards and processes operate from the machine age perspective and have not shifted to the more holistic, systems-age model of system safety. A system-based accident model that invokes a systems-theoretic view of causality called STAMP is worthy of review.

8.5 A Systems Based Accident Model

The Systems-Theoretic Accident Model and Processes (STAMP) changes the machine-age emphasis from preventing failures to the systems-age concept of enforcing safety constraints (Leveson 2004; Leveson et al. 2009). The next two sections will review the principles that support STAMP and how it intersects with systems design endeavors.

8.5.1 Systems-Theoretic Principles of STAMP

STAMP is a systems-theoretic accident model because it invokes system theory's centrality axiom and its pair of supporting principles—hierarchy and emergence and communications and control to enforce safety constraints in both systems designs and subsequent operations.

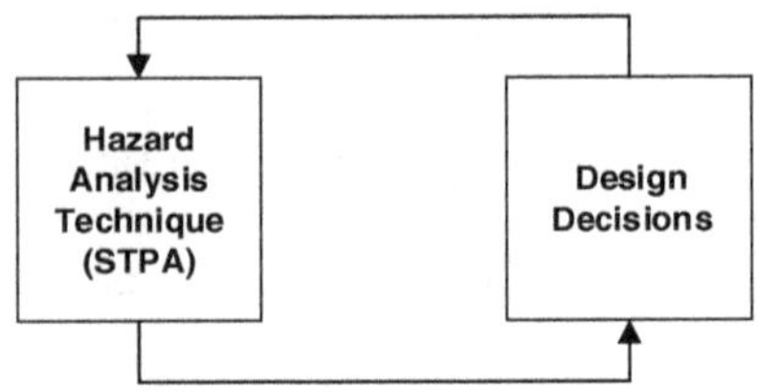

Fig. 8.1 Safety-guided design [based on Fig. 9.1 in Leveson (2011)]

- "Safety is an emergent property of systems that arises from the interaction of system components" (Leveson 2004, p. 249).
- A hierarchy of control structures are assembled purposefully. "Control processes operate between levels to control to control the processes at lower levels in the hierarchy. These control processes enforce the safety constraints for which the control process is responsible" (Leveson 2011, p. 81).
- Control is how system safety constraints are enforced and may be either passive or active.
- Communication through the use of control feedback loops keeps the system in a state of dynamic equilibrium.

STAMP uses an accident scenario identification technique that includes hazards caused by: (1) design errors; (2) component interaction accidents; (3) cognitively complex human decision-making errors; and (4) social, organizational, and management factors contributing to accidents (Leveson 2011). The technique is called System-Theoretic Process Analysis (STPA).

STPA is integrated with the system design process in what is termed safety-guided design, depicted in Fig. 8.1.

The next section will discuss how STAMP is integrated within the system design process.

8.5.2 Intersection of STAMP Criteria and Systems Design

The development of adequate criteria for evaluating systems safety requires an evaluation of the systems-theoretic principles of STAMP against the safety related design activities and tasks from *IEEE Standard 1220—Systems engineering—Application and management of the systems engineering process* (IEEE 2005) that were described in the section on safety in system design efforts. The criteria intersection is presented in Table 8.3.

In summary, STAMP may be described as a systems-theoretic model for safety that:

Table 8.3 Intersection of STAMP criteria and systems design

Design process	IEEE Standard 1220 section and task	STAMP criteria
Requirements analysis	6.1.1—Stakeholder expectations are balanced with an analysis of the effects on the overall system design and safety	System-level safety constraints
	6.1.4—Measures of effectiveness are defined to reflect overall stakeholder expectations and satisfaction that include safety	System-level safety constraints
	6.1.9.5—The design team accounts for the system design features that create significant hazards	System-theoretic process analysis (STPA)
Functional analysis	6.3.2.5 (1)—The design team analyzes and prioritizes potential functional failure modes to define failure effects and identify the need for fault detection and recovery functions	System control structure responsible for enforcing safety constraints identified by the STPA
	6.3.2.5 (2)—Functional reliability models are established to support the analysis of system effectiveness for each operational scenario	
	6.3.2.5 (3)—Failures, which represent significant safety hazards, are modeled to completely understand system impacts	
	6.3.2.6 (1)—The design team analyzes subfunctions to identify operational hazards	(1) System-theoretic process analysis (STPA)
	6.3.2.6 (2)—Additional functional requirements are derived and defined for monitoring dangerous operational conditions, or notifying or warning operators of impending hazards	(2) System control structure responsible for enforcing safety constraints identified by the STPA
Synthesis	6.5.3—The design team analyzes all design to identify potential hazards to the system, humans involved in the system and supporting the system life cycle processes, or the environment	System-theoretic process analysis (STPA)
Systems analysis	6.7.6.3—The design team identifies safety impacts associated with system implementation. Safety laws and regulations should be identified, and the design team should ensure that these are complied with by each solution alternative	System-theoretic process analysis (STPA)

> STAMP focuses particular attention on the role of constraints in safety management. Instead of defining safety in terms of preventing component failure events, it is defined as a continuous control task to impose the constraints necessary to limit system behavior to safe changes and adaptations. Accidents are seen as resulting from inadequate control or enforcement of constraints on safety-related behavior at each level of the system development and system operations control structures. Accidents can be understood, therefore, in terms of why the controls that were in place did not prevent or detect maladaptive changes, that is, by identifying the safety constraints that were violated at each level of the control structure as well as why the constraints were inadequate or, if they were potentially adequate, why the system was unable to exert appropriate control over their enforcement. The process leading to an accident (loss event) can be described in terms of an adaptive feedback function that fails to maintain safety as performance changes over time to meet a complex set of goals and values. The adaptive feedback mechanism allows the model to incorporate adaptation as a fundamental property. (Leveson 2004, pp. 265–266)

While an in-depth review of each STAMP is beyond this chapter, readers are encouraged to consult Part III—Using STAMP in Leveson's (2011) text *Engineering a Safer World: Systems Thinking Applied to Safety.*

The next section will discuss a measure for evaluating system safety.

8.6 A Measure for Evaluating System Safety

In the previous sections the use of a systems-based model for ensuring that system safety was advocated and emerges as a purposeful result of the design process. In order to ensure that the system design process has invoked a holistic, socio-technical perspective the design effort should be evaluated using the essential criteria of such a model. As with traceability, the criteria will be subjective, qualitative measures that will need to answer questions that address both the presence (yes or no) and quality of the effort (how well) to provide a sufficiently robust systems-based model as an element of the system design process. In this case the STAMP criteria will need to be related to a specific measurable attribute that can be utilized as a measure. Once again, measures are important because they are the linkage between observable, real-world, empirical facts and the construct (i.e., system safety model) that we create as an evaluation point.

8.6.1 Scale for System Safety

As discussed during the development of scales for previous non-functional requirements, the selection of a measurement scale is an important element in the development of an adequate measure for system safety. Because the STAMP criteria selected for safety have no natural origin or empirically defined distance, the ordinal scale was selected as an appropriate scale for measuring the safety attributes. The well-known Likert scale is proposed for use in evaluating system safety. In order to improve reliability a five point Likert scale will be invoked (Lissitz and Green 1975).

Before moving on to describing the measure for system safety two important points must once again be made with respect to scale development. Scales are characterized as either a *proposed* scale or a scale. "A proposed scale is one that some investigator(s) put forward as having the requisite properties, and if it is indeed shown to have them, then it is recognized as a scale" (Cliff 1993, p. 65). In this chapter use of the word scale is referring to *proposed scales*. As stated before, this may seem to be an insignificant point, but until the scale has been accepted and successfully utilized it remains proposed.

8.6.2 Proposed Measurement Scale for System Safety

Armed with a construct, measurement attributes and an appropriate scale type, the system safety measure may be constructed. In order to evaluate system safety, the need to answer questions that address both the presence (yes or no) and quality of the effort (how well) to provide system safety by invoking the principal elements of a systems-based model during systems design endeavors must be answered. The seven STAMP criteria (i.e., our measurement constructs) from Table 8.3 have been rearranged in Table 8.4 and in order to evaluate the design's ability to conform to the STAMP criteria for system safety, a specific question has been developed which may be used to evaluate each of the seven system safety measurement concerns.
The answers to the questions will be contained in a 5 point-Likert scale. The measurement constructs and questions associated with each of the measurements concerns are presented in Table 8.5.
The answer to each question will be scored using the 5-point Likert measures in Table 8.6
A generalized measure for system safety is shown in Eq. 8.1.

Generalized Equation for Systems Safety

$$S_{sys} = \sum_{i=1}^{n} S_i \tag{8.1}$$

The overall measure for system safety is a sum of the scores from the seven system safety metrics as shown in Eq. 8.2 and will be the measure the degree of system safety in a system design endeavor.

Expanded Equation for System Safety

$$S_{sys} = S_{ra1} + S_{ra2} + S_{fa1} + S_{fa2} + S_{fa3} + S_{syn} + S_{sa} \tag{8.2}$$

The next section will discuss how to measure safety.

Table 8.4 Measurement constructs for system safety

Design process and IEEE Standard 1220 section		STAMP criteria	System safety measurement concern
Requirements analysis	6.1.1	System-level safety constraints	1. Does the requirements analysis process include an analysis of system-level safety constraints?
	6.1.4	System-level safety constraints	
	6.1.9.5	System-theoretic process analysis (STPA)	2. Does the requirements analysis process include system-theoretic process analysis (STPA)?
Functional analysis	6.3.2.5	System control structure responsible for enforcing safety constraints identified by the STPA	3. Does the functional analysis process develop control structures responsible for enforcing safety constraints identified by the STPA?
	6.3.2.6 (1)	(1) System-theoretic process analysis (STPA)	4. Does the functional analysis process include system-theoretic process analysis (STPA)?
	6.3.2.6 (2)	(2) System control structure responsible for enforcing safety constraints identified by the STPA	5. Does the functional analysis process develop control structures responsible for enforcing safety constraints identified by the STPA?
Synthesis	6.5.3	System-theoretic process analysis (STPA)	6. Does the synthesis process include system-theoretic process analysis (STPA)?
Systems analysis	6.7.6.3	System-theoretic process analysis (STPA)	7. Does the systems analysis process include system-theoretic process analysis (STPA)?

8.7 Measuring System Safety

At the end of Chap. 3 the importance of being able to measure each non-functional attribute was stressed as an essential element in systems design endeavors. A structural mapping that relates system safety to a specific metric and measurement entity are required. The four-level construct for system safety is presented in Table 8.7.

Table 8.5 Measurement questions for design traceability

Measurement construct	Traceability concern for measurement
S_{ra1}	Does the requirements analysis process include an analysis of system-level safety constraints?
S_{ra2}	Does the requirements analysis process include system-theoretic process analysis (STPA)?
S_{fa1}	Does the functional analysis process develop control structures responsible for enforcing safety constraints identified by the STPA?
S_{fa2}	Does the functional analysis process include system-theoretic process analysis (STPA)?
S_{fa3}	Does the functional analysis process develop control structures responsible for enforcing safety constraints identified by the STPA?
S_{syn}	Does the synthesis process include system-theoretic process analysis (STPA)?
S_{sa}	Does the systems analysis process include system-theoretic process analysis (STPA)?

Table 8.6 Traceability measurement question Likert scale

Measure	Descriptor	Measurement criteria
0.0	None	No objective quality evidence is present
0.5	Limited	Limited objective quality evidence is present
1.0	Nominal	Nominal objective quality evidence is present
1.5	Wide	Wide objective quality evidence is present
2.0	Extensive	Extensive objective quality evidence is present

Table 8.7 Four-level structural map for measuring system safety

Level	Role
Concern	System safety
Attribute	Safety
Metric	System safety
Measurable characteristic	Safety of (1) requirements analysis process (S_{ra1}, S_{ra2}), (2) functional analysis process (S_{fa1}, S_{fa2}, S_{fa3}), (3) synthesis process (S_{syn}), and (4) systems analysis (S_{sa})

8.8 Summary

In this chapter the non-functional requirement for safety has been reviewed. A formal definition for safety has been provided along with additional explanatory definitions, terms, and equations. The ability to effect safety during the design process has also been addressed. Finally, a formal metric and measurement characteristic have been proposed for evaluating the non-functional requirement for safety.

The next Part of the text will shift the focus to adaptation concerns. Adaptation concerns address the system's ability to change and adapt in order to remain viable and continue to address the requirements of its stakeholders. The first chapter in the Part on Adaptation Concerns will address the non-functional attributes for adaptability, flexibility, modifiability and flexibility, and robustness. The second chapter in Part III on Adaptation Concerns will address the non-functional attributes for extensibility, portability, reusability, and self-descriptiveness.

References

Adams, K. M., Hester, P. T., Bradley, J. M., Meyers, T. J., & Keating, C. B. (2014). Systems theory: The foundation for understanding systems. *Systems Engineering, 17*(1), 112–123.

Alberico, D., Bozarth, J., Brown, M., Gill, J., Mattern, S., & McKinlay, A. (1999). *Software system safety handbook: A technical and managerial team approach.* Washington: Joint Services Software Safety Committee.

Cliff, N. (1993). What is and isn't measurement. In G. Keren & C. Lewis (Eds.), *A handbook for data analysis in the behavioral sciences: Methodological issues* (pp. 59–93). Hillsdale: Lawrence Erlbaum Associates.

DoD. (2000). *Military Standard (MIL-STD-882D): Standard practice for system safety.* Washington: Department of Defense.

IEEE. (1994). *IEEE Standard 1228: Software safety plans.* New York: Institute of Electrical and Electronics Engineers.

IEEE. (2005). *IEEE Standard 1220: Systems engineering—application and management of the systems engineering process.* New York: Institute of Electrical and Electronics Engineers.

IEEE, & ISO/IEC. (2010). *IEEE and ISO/IEC Standard 24765: Systems and software engineering—vocabulary.* New York, Geneva: Institute of Electrical and Electronics Engineers and the International Organization for Standardization and the International Electrotechnical Commission.

Leveson, N. G. (2004). A new accident model for engineering safer systems. *Safety Science, 42*(4), 237–270.

Leveson, N. G. (2011). *Engineering a safer world: Systems thinking applied to safety.* Cambridge: MIT Press.

Leveson, N. G., Dulac, N., Marais, K., & Carroll, J. (2009). Moving beyond normal accidents and high reliability organizations: A systems approach to safety in complex systems. *Organization Studies, 30*(2–3), 227–249.

Lissitz, R. W., & Green, S. B. (1975). Effect of the number of scale points on reliability: A Monte Carlo approach. *Journal of Applied Psychology, 60*(1), 10–13.

NASA. (2011). *NASA System Safety Handbook (NASA/SP-2010-580).* In System Safety Framework and Concepts for Implementation (Vol. 1). Washington: National Aeronautics and Space Administration.

Parker, S. (Ed.). (1994). *McGraw-Hill dictionary of engineering.* New York: McGraw-Hill.

Penzenstadler, B., Raturi, A., Richardson, D., & Tomlinson, B. (2014). Safety, security, now sustainability: The nonfunctional requirement for the 21st century. *IEEE Software, 31*(3), 40–47.

Perrow, C. (1999). *Normal accidents: Living with high-risk technologies.* Princeton: Princeton University Press.

Part IV
Adaptation Concerns

Chapter 9
Adaptability, Flexibility, Modifiability and Scalability, and Robustness

Abstract The design of systems and components during the design stage of the systems life cycle requires specific purposeful actions to ensure effective designs and viable systems. Designers are faced with a number of *adaptation concerns* that they must embed into the design in every instance of thinking and documentation. The ability for a systems to change is essential to its continued survival and ability to provide requisite functions for its stakeholders. Changeability includes the non-functional requirements for adaptability, flexibility, modifiability and robustness. Purposeful design requires an understanding of each of these requirements and how to measure and evaluate each as part of an integrated systems design.

9.1 Introduction to Changeability

This chapter will address four major topics: (1) adaptability; (2) flexibility; (3) modifiability and scalability; and (4) robustness in design endeavors. The chapter begins with a section that reviews the concept of changeability, its three unique elements, and a method for representing systems change using a state-transition-diagram.

Section 9.2 defines adaptability and flexibility and provides a clear method for distinguishing between these two non-functional properties.

Section 9.3 in this chapter addresses modifiability by providing a clear definition and a distinction between it and both scalability and maintainability.

Section 9.4 defines robustness and discusses the design considerations related to robust systems.

Section 9.5 defines a measure and a means for measuring changeability that is a function of (1) adaptability; (2) flexibility; (3) modifiability; and (4) robustness. The chapter completes by relating the proposed measure for changeability as a metric and includes a structural map for traceability.

The chapter has a specific learning goal and associated objectives. The learning goal of this chapter is to be able to identify how the attributes of adaptability,

K.M. Adams, *Non-functional Requirements in Systems Analysis and Design*,
Topics in Safety, Risk, Reliability and Quality 28,
DOI 10.1007/978-3-319-18344-2_9

flexibility, modifiability and scalability, and robustness that influence design in systems endeavors. This chapter's goal is supported by the following objectives:

- Describe changeability using a state-transition-diagram.
- Define adaptability.
- Define flexibility.
- Describe the difference between adaptability and flexibility.
- Define modifiability.
- Describe the difference between modifiability and maintainability.
- Define robustness.
- Describe the design factors that determine robustness.
- Construct a structural map that relate changeability to a specific metric and measureable characteristic.
- Explain the significance of adaptability, flexibility. Modifiability, and robustness in systems design endeavors.

The ability to achieve these objectives may be fulfilled by reviewing the materials in the sections that follow.

9.2 The Concept of Changeability

The motivation for change in existing systems is based upon three major factors: (1) marketplace forces; (2) technological evolution; and (3) environmental shifts (Fricke and Schulz 2005). These drivers of change must be addressed by systems practitioners throughout the system lifecycle. Because real-world systems exist in a constantly changing domain, they too are subject to change. The ability for a system to change is termed *changeability*. Changeability is a term that is not formally defined in the systems engineering vocabulary, but it encompasses a number of defined terms that include adaptability, flexibility, modifiability, scalability, and robustness. Each of these terms will be fully addressed in later sections.

Right now, the most important point is that changeability addresses differences in a system over time. Change can be thought of simply as the differences in a system between and initial or zero time t_o and time t_f where f = some future time. During the time transition between t_o and t_f, either the system or environment or both the systems and environment may have been altered. A systems' life cycle is filled with changes to both the system and its related environment. The constant stream of changes that occur during the systems life cycle require the system's designers and maintainers to plan for, recognize, and control changes to ensure the system remains both viable and functionally effective. Changes that occur in a system are characterized by a change event that contains three unique elements: (1) the cause or impetus for the change; (2) the mechanism that affects the change, and (3) the overall effect of change on both the system and its environment. These events will be discussed in the sections that follow.

9.2.1 Agent for Change

The impetus for change originates as a result of one or more of three change factors described in the previous section: (1) marketplace forces; (2) technological evolution; and (3) environmental shifts (Fricke and Schulz 2005). The source, which is the instigator, force, or impetus that affects the change is labeled the *change agent*. The change agent is responsible for transforming the change factor into action (Ross et al. 2008).

9.2.2 Mechanism of Change

The mechanism of change describes the path taken between time t_o and time t_f while the system and or its environment is being transformed from its initial state to the new altered state (Ross et al. 2008). The pathway is the action that includes all of the mechanistic resources (i.e., material, manpower, money, minutes (time), methods, and information) required to affect the change.

9.2.3 Effects of Change on Systems and Their Environment

The effects of change are the actual differences between the system and or the environment at t_o and time t_f (Ross et al. 2008). The difference between the system at t_o and t_f is described in terms of the changed or newly added characteristics (y_i) which are the results of a series of discrete events put into motion by the change agent and accomplished by specific mechanisms (m_i). Each mechanism may be depicted as a transition arc with a discrete event and subsequent action that cause a change in system characteristics (Y where $y_i \in Y$).

9.2.4 Depicting Change Events

The time-dependent behavior of the system and its environment, as a function of the three change elements just described, may be modeled in a state transition diagram (Hatley and Pirbhai 1988). The state-transition-diagram (STD) accounts for the impetus, agent, pathways, and effects during the change event. Figure 9.1 is an STD that defines change as a function of the change agent, events and actions, mechanisms and effects where:

- Y = A set of n system characteristics Y_i, where i = 1 to n and $y_i \in Y$.
- y_i = A system characteristic y_i, where i = 1 to n.

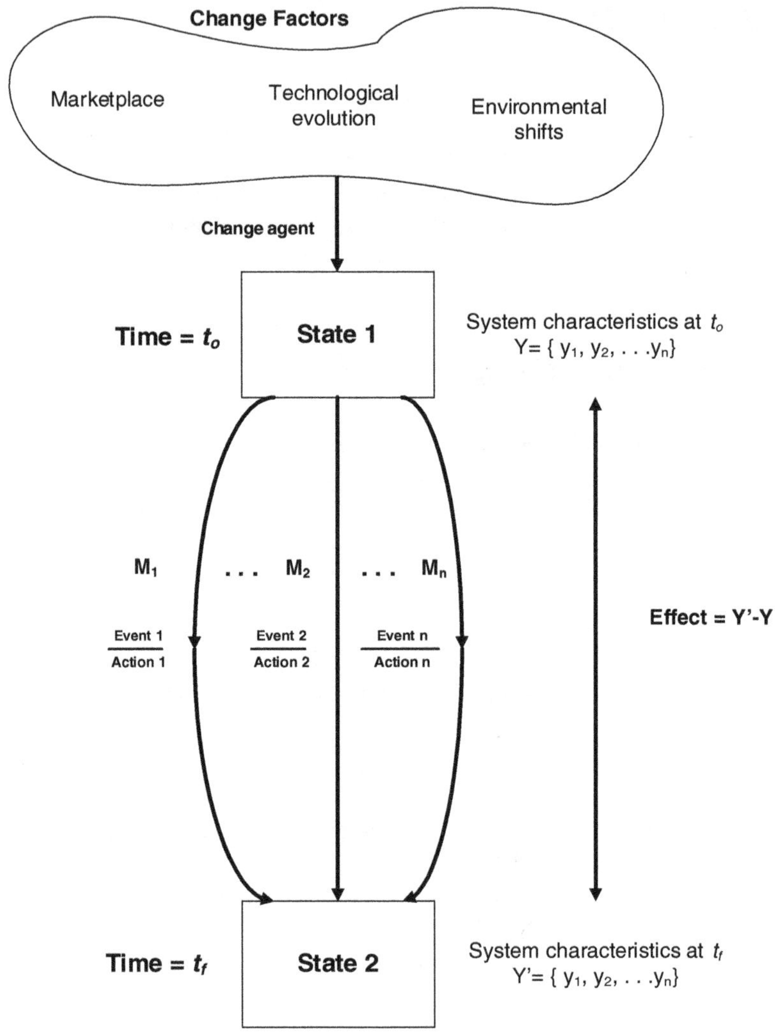

Fig. 9.1 State transition diagram for system change

- M = A set of n mechanisms m_i, where $i = 1$ to n and $m_i \in M$.
- m_i = The mechanisms m_i, where $i = 1$ to n.

The transition arcs and arrows that connect the state boxes show the event or events required to cause the change of state, and the resultant action when the imposed change agent is applied.

Armed with a basic understanding of change and the concepts that surround changeability a discussion about the five non-functional properties: (1) adaptability; (2) flexibility; (3) modifiability; (4) scalability; and (5) robustness and how they are applied during systems endeavors.

9.3 Adaptability and Flexibility

In this section the basics of adaptability and flexibility and how they are applied during systems endeavors are discussed. Adaptability and flexibility have many interpretations and as such must be clearly defined and understood.

9.3.1 Adaptability Definition

Adaptability, from a systems engineering perspective, is defined as:

> Degree to which a product or system can effectively and efficiently be adapted for different or evolving hardware, software or other operational or usage environments [SE VOCAB].

Adaptability has additional definitions, shown in Table 9.1 that may provide improved understanding of the term when used as a non-functional requirement for a system.

The section that follows will provide the definition for flexibility.

9.3.2 Flexibility Definition

Flexibility, from a systems engineering perspective, is defined as:

> The ease with which a system or component can be modified for use in applications or environments other than those for which it was specifically designed. Syn: adaptability cf. extendability, maintainability (IEEE & ISO/IEC 2010, p. 144).

Table 9.1 Additional definitions for adaptability

Definition	Source
"A characteristic of a system amenable to change to fit altered circumstances, where "circumstances" include both the context of a system's use and its stakeholders' desires"	Engel and Browning (2008, p. 126)
"The ability to change (e.g., to improve performance over a period of time) within a given state"	Bordoloi et al. (1999, p. 135)
"Characterizes a system's ability to adapt itself towards changing environments. Adaptable systems deliver their intended functionality under varying operating conditions through changing themselves. That is no changes from external have to be implemented into such systems to cope with changing environments"	Fricke and Schulz (2005, p. 347)
"The degree to which adjustments in practices, processes, or structures of systems are possible to projected or actual changes of its environment"	Andrzejak et al. (2006, p. 30)

Table 9.2 Additional definitions for flexibility

Definition	Source
"The ability to change "states"	Bordoloi et al. (1999, p. 135)
"Characterizes a system's ability to be changed easily. Changes from external have to be implemented to cope with changing environments"	Fricke and Schulz (2005, p. 347)

Flexibility has additional definitions, shown in Table 9.2 that may provide improved understanding of the term when used as a non-functional requirement for a system.

The section that follows will show how adaptability and flexibility are related.

9.3.3 Relationship Between Adaptability and Flexibility

A taxonomic distinction between adaptability and flexibility may be made based upon the location of the source causing the system to change. The source (e.g., instigator, force, or impetus) of the change has been termed the *change agent* and the location vis-à-vis the system serves to make the distinction between adaptability and flexibility (Ross et al. 2008). Figure 9.2 depicts both an internal and an external change agent acting on a system and the resulting classification as either adaptability or flexibility.

Based upon the definitions for adaptability and flexibility provided in Tables 9.1 and 9.2, and the relationship depicted in Fig. 9.2, the relationship between adaptability and flexibility is as follows:

- *Adaptability* An internal impetus for change causes a system to change within a given state, which is termed adaptable change.
- *Flexibility* An external impetus for changes causes a system to change states, which is termed flexible change. A system *S* is said to changes states to system *S'* when S', as a result of change is now able "… to produce new goods or services, or to produce the present array of goods and services in ways or volumes that are not possible in the former state" (Bordoloi et al. 1999, p. 135).

The next section will discuss modifiability.

9.4 Modifiability and Scalability

In this section the basics of modifiability and how it is applied during systems endeavors are reviewed. Modifiability has been interpreted in many ways and as such must be clearly defined and understood.

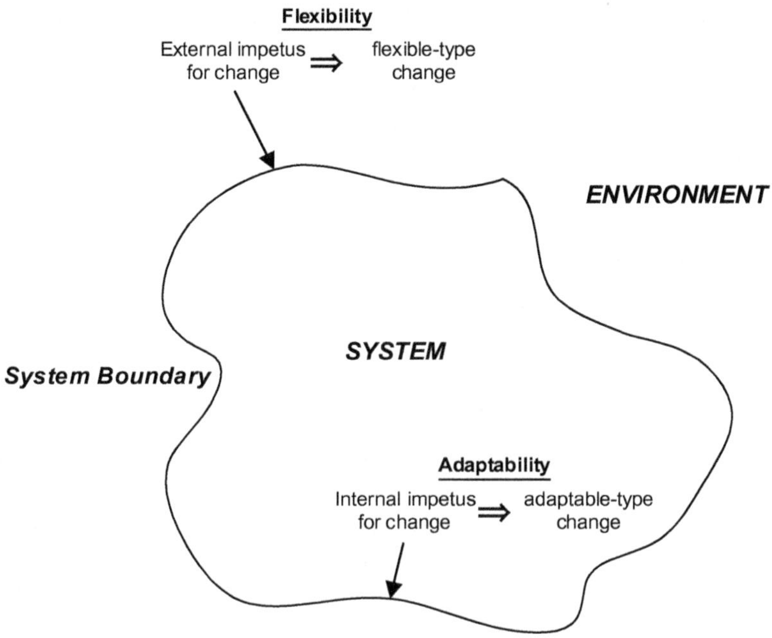

Fig. 9.2 Change agent location in distinguishing between adaptability and flexibility

9.4.1 Modifiability Definition

Modifiability, from a systems engineering perspective, is defined as:

> The ease with which a system can be changed without introducing defects cf. maintainability (IEEE & ISO/IEC 2010, p. 222).

Modifiability has additional definitions, shown in Table 9.3 that may provide improved understanding of the term when used as a non-functional requirement for a system.

It is important to note that modifiability is interested in the set of system characteristics where:

- Y = A set of n system characteristics Y_i, where i = 1 to n and $y_i \in Y$.
- y_i = A system characteristic y_i, where i = 1 to n.

Table 9.3 Additional definitions for modifiability

Definition	Source
"The ability to change the membership of the parameter set"	Ross et al. (2008, p. 249)
"The ease with which it can be modified to changes in the environment, requirements or functional specification"	Bengtsson et al. (2004, p. 130)

There are two important distinctions to make when considering the definition for modifiability.

1. The magnitude or level of the individual characteristics is addressed by *scalability* and is not a concern for modifiability since no new characteristics are being introduced into the set of system characteristics (*Y*). [NOTE: *Scalability* will no longer be discussed]
2. The essential difference between maintainability and modifiability, is that maintainability is concerned with the correction of bugs whereas modifiability is not.

9.4.2 Modifiability in Systems

Modifiability, from a systems design perspectives is related to modularity. In review, modularity was defined as "the degree to which a system or computer program is composed of discrete components such that a change to one component has minimal impact on other components" (IEEE & ISO/IEC 2010, p. 223). Based upon this definition for modularity, it is reasonable to assume that a system that has a high degree of modularity (i.e., independent system components) would be more easily modified. This is because a change in an independent module is easier to accomplish than one that is closely or tightly coupled to many other modules.

Modularity is a characteristics of a good system design. By designing system components with a single purpose and clearly defined functions, inputs, and outputs change is more easily affected. As stated earlier, engineering designs with high modularity accomplish many things, which include:

- *First, it makes the complexity of the system manageable by providing an effective "division of cognitive labor."*
- *Second, modularity organizes and enable parallel work.*
- *Finally, modularity in the 'design' of a complex system allows modules to be changed and improved over time without undercutting the functionality of the system as a whole* (Baldwin and Clark 2006, p. 180).

In summary, a highly modular system with have improved modifiability.

The next section will discuss robustness.

9.5 Robustness

In this section the basics of robustness and how it is applied during systems endeavors is reviewed. As with most of the non-functional requirements, robustness has been interpreted in many ways and as such must be clearly defined and understood.

9.5.1 Robustness Definition

Robustness, from a systems engineering perspective, is defined as:

> The degree to which a system or component can function correctly in the presence of invalid inputs or stressful environmental conditions cf. error tolerance, fault tolerance (IEEE & ISO/IEC 2010, p. 313).

Robustness has additional definitions, shown in Table 9.4 that may provide improved understanding of the term when used as a non-functional requirement for a system.

9.5.2 Robustness in Systems

It is important to note that robustness is often referring to the larger system, without addressing any particular system characteristic, individual component, subsystem, or environmental perturbation. In fact, robustness is thought to be a function of a system's internal structure and fragility.

> Because robustness is achieved by very specific internal structures, when any of these systems is disassembled, there is very little latitude in reassembly if a working system is expected. Although large variations or even failures in components can be tolerated if they are designed for through redundancy and feedback regulation, what is rarely tolerated, because it is rarely a design requirement, is nontrivial rearrangements of the interconnection of internal parts (Carlson and Doyle 2002, p. 2539).

The study of complex systems has provided a framework titled Highly Optimized Tolerance (HOT) that seeks to focus attention on robustness through both tolerance and configuration.

> 'Tolerance' emphasizes that robustness in complex systems is a constrained and limited quantity that must be carefully managed and protected. 'Highly optimized' emphasizes that this is achieved by highly structured, rare, non-generic configurations that are products either of deliberate design or evolution. The characteristics of HOT systems are high

Table 9.4 Additional definitions for robustness

Definition	Source
"The maintenance of some desired system characteristic despite fluctuations in the behavior of its component parts or its environment"	Carlson and Doyle (2002, p. 2539)
"The ability to remain "constant" in parameters in spite of system internal and external changes"	Ross et al. (2008, p. 249)
"Characterizes a systems ability to be insensitive towards changing environments. Robust systems deliver their intended functionality under varying operating conditions without being changed"	Fricke and Schulz (2005, p. 347)

> performance, highly structured internal complexity, and apparently simple and robust external behavior, with the risk of hopefully rare but potentially catastrophic cascading failure events initiated by possibly quite small perturbations (Carlson and Doyle 2002, p. 2540).

Robustness as a non-functional requirement for a system is used to evaluate a system's ability to maintain constant design parameters in the face of either individual or simultaneous internal or external perturbations. Systems designers are able to influence robustness by carefully constructing complex systems with purposeful internal component configuration and redundancy that accomplish both function and provide high levels of performance.

9.6 Changeability in Systems Design Efforts

The ability to understand, measure, and evaluate the changeability of a system would seem to be a valuable capability. Understanding changeability and its constituent non-functional properties (adaptability, flexibility, modifiability, and robustness) and having the ability to measure and evaluate it provides additional perspectives and insight into the future performance and viability of all elements of the system being designed. The formal design process should address each of element of changeability.

In the traditional systems design process invoked by *IEEE Standard 1220—Systems engineering—Application and management of the systems engineering process* (IEEE 2005) neither changeability nor any of its component non-functional characteristics (adaptability, flexibility, modifiability, and robustness) are mentioned. Despite this, changeability is an important aspect to be considered in the conceptual, preliminary, and detailed design stages of the systems lifecycle.

The section that follows will discuss changeability may be evaluated as a function of its four non-functional requirements: (1) adaptability; (2) flexibility; (3) modifiability; and (4) robustness.

9.6.1 A Method for Evaluating Changeability

Based on the understanding developed with respect to understanding changeability presented in the previous sections and how it may be used to evaluate systems characteristics it is time to develop an appropriate measure. Measuring something like changeability is tough, because changeability is a subjective, qualitative measure which differs from most of the objective, quantitative measures developed for the non-functional requirements addressed so far. In order to understand how to approach a subjective, qualitative measure, a review of how to construct and measure subjective, qualitative objects is in order.

9.6.1.1 Development of Measurement Scales

In order to evaluate changeability, questions that address both the presence (yes or no) and quality of the effort (how well) to provide changeability as a purposeful effort during a system design endeavor must be developed and answered. In this case each of the four non-functional requirements identified as constituting the measure termed changeability must be addressed. The goal is to frame each of the four non-functional requirements as an object with a specific measureable attribute. The establishment of measures is important because they are the linkage between the observable, real-world, empirical facts about the system and the construct (i.e., changeability) devised as an evaluation point. In this case a measure is defined as "... an observed score gathered through self-report, interview, observation, or some other means" (Edwards and Bagozzi 2000, p. 156).

9.6.1.2 Scale for Changeability

As we discussed during the development of the scales for both traceability (see Sect. 5.4) and system safety (see Sect. 6.4.5), the selection of a measurement scale is an important element in the development of an adequate measure for changeability. Because none of the non-functional requirements we have selected as criteria for changeability have a natural origin or empirically defined distance, an ordinal scale should be selected as an appropriate scale for measuring system changeability. The well-known Likert scale is proposed for use in evaluating changeability. In order to ensure improved reliability a five point Likert scale will be invoked (Lissitz and Green 1975).

9.6.1.3 Proposed Scales

Before moving on to describing the measure for changeability an important point with respect to scale development must be stated once again. Scales are characterized as either a *proposed* scale or a scale. "A proposed scale is one that some investigator(s) put forward as having the requisite properties, and if it is indeed shown to have them, then it is recognized as a scale" (Cliff 1993, p. 65). As previously stated, the use of the word scale is referring to *proposed scales*. This may seem to be an insignificant point, but until the scale has been accepted and successfully utilized it remains proposed.

9.6.1.4 Proposed Measurement Scale for Changeability

Armed with a construct, measurement attributes and an appropriate scale type, the changeability measure may be constructed. In order to evaluate changeability, questions that address both the presence (yes or no) and quality of the effort (how well) must be answered to provide a measure for changeability. This is accomplished by invoking each of the four non-functional requirements for changeability: (1) adaptability; (2) flexibility; (3) modifiability; and (4) robustness. Each of the four changeability criteria (i.e., the measurement constructs) have a specific question, shown in Table 9.5, which may be used to evaluate their contribution to changeability.

The answer to each question in Table 9.5 will be scored using the 5-point Likert measures in Table 9.6.

The summation of the four (4) constructs in Eq. 9.1 will be the measure the degree of changeability in a system design endeavor.

Expanded Equation for System Changeability

$$Ch_{sys} = Ch_{adapt} + Ch_{flex} + Ch_{modif} + Ch_{robust} \tag{9.1}$$

9.6.2 Measuring Changeability

At the end of chapter 3 the importance of being able to measure each non-functional attribute was highlighted as being an important element of systems design. A structural mapping that relates changeability to a specific metric and measurement entity are required. The four-level construct for changeability is presented in Table 9.7.

Table 9.5 Measurement questions for changeability

Measurement construct	Changeability concern for measurement
Ch_{adapt}	Is the system able to adapt itself as a result of states changes caused by internal impetus?
Ch_{flex}	Is the system flexible enough to change as a result of state changes caused by external environmental impetus?
Ch_{modif}	Can the system be modified as a result of changes in the environment, requirements or functional specification?
Ch_{robust}	Can the system's parameters remain "constant" in spite of system internal or external environmental changes?

Table 9.6 Changeability measurement question Likert scale

Measure	Descriptor	Measurement criteria
0.0	None	No objective quality evidence is present
0.5	Limited	Limited objective quality evidence is present
1.0	Nominal	Nominal objective quality evidence is present
1.5	Wide	Wide objective quality evidence is present
2.0	Extensive	Extensive objective quality evidence is present

Table 9.7 Four-level structural map for measuring changeability

Level	Role
Concern	Systems adaptation
Attribute	Changeability
Metric	System changeability
Measurable characteristic	Changeability of (1) adaptability (Ch_{adapt}), (2) flexibility (Ch_{flex}), (3) modifiability (C_{modif}), and (4) robustness (Ch_{robust})

9.7 Summary

This chapter has addressed the adaptation concern for changeability and reviewed its four non-functional requirements: (1) adaptability; (2) flexibility; (3) modifiability; and (4) robustness. In each case a formal definition has been provided along with additional explanatory definitions and terms. The ability to effect the non-functional requirement during the design process has also been addressed. Finally, a formal metric and measurement characteristic have been proposed for evaluating changeability.

The chapter that follows will address non-functional requirement for extensibility, portability, reusability, and self-descriptiveness as part of the concern for adaptation in systems endeavors.

References

Andrzejak, A., Reinefeld, A., Schintke, F., & Schütt, T. (2006). On adaptability in grid systems. In V. Getov, D. Laforenza, & A. Reinefeld (Eds.), *Future generation grids* (pp. 29–46). New York, US: Springer.

Baldwin, C. Y., & Clark, K. B. (2006). Modularity in the design of complex engineering systems. In D. Braha, A. A. Minai, & Y. Bar-Yam (Eds.), *Complex engineered systems* (pp. 175–205). Berlin: Springer.

Bengtsson, P., Lassing, N., Bosch, J., & van Vliet, H. (2004). Architecture-level modifiability analysis (ALMA). *Journal of Systems and Software, 69*(1–2), 129–147.

Bordoloi, S. K., Cooper, W. W., & Matsuo, H. (1999). Flexibility, adaptability, and efficiency in manufacturing systems. *Production and Operations Management, 8*(2), 133–150.

Carlson, J. M., & Doyle, J. (2002). Complexity and robustness. *Proceedings of the National Academy of Sciences of the United States of America, 99*(3), 2538–2545.

Cliff, N. (1993). What Is and Isn't Measurement. In G. Keren & C. Lewis (Eds.), *A handbook for data analysis in the behavioral sciences: Methodological issues* (pp. 59–93). Hillsdale, NJ: Lawrence Erlbaum Associates.

Edwards, J. R., & Bagozzi, R. P. (2000). On the nature and direction of relationships between constructs and measures. *Psychological Methods, 5*(2), 155–174.

Engel, A., & Browning, T. R. (2008). Designing systems for adaptability by means of architecture options. *Systems Engineering, 11*(2), 125–146.

Fricke, E., & Schulz, A. P. (2005). Design for changeability (DfC): Principles to enable changes in systems throughout their entire lifecycle. *Systems Engineering, 8*(4), 342–359.

Hatley, D. J., & Pirbhai, I. A. (1988). *Strategies for real-time system specification*. New York: Dorset House.

IEEE. (2005). *IEEE Standard 1220: Systems engineering—Application and management of the systems engineering process*. New York: Institute of Electrical and Electronics Engineers.

IEEE, & ISO/IEC. (2010). *IEEE and ISO/IEC Standard 24765: Systems and software engineering —Vocabulary*. New York and Geneva: Institute of Electrical and Electronics Engineers and the International Organization for Standardization and the International Electrotechnical Commission.

Lissitz, R. W., & Green, S. B. (1975). Effect of the number of scale points on reliability: A Monte Carlo approach. *Journal of Applied Psychology, 60*(1), 10–13.

Ross, A. M., Rhodes, D. H., & Hastings, D. E. (2008). Defining changeability: Reconciling flexibility, adaptability, scalability, modifiability, and robustness for maintaining system lifecycle value. *Systems Engineering, 11*(3), 246–262.

Chapter 10
Extensibility, Portability, Reusability and Self-descriptiveness

Abstract The design of systems and components during the design stage of the systems life cycle requires specific purposeful actions to ensure effective designs and viable systems. Designers are faced with a number of *adaptation concerns* that they must embed into the design in every instance of thinking and documentation. The ability for a system to adapt is essential to its continued survival and ability to provide requisite functions for its stakeholders. Adaptation concerns includes the non-functional requirements for extensibility, portability, reusability, and self-descriptiveness. Purposeful design requires an understanding of each of these requirements and how to measure and evaluate each as part of an integrated systems design.

10.1 Introduction to Extensibility, Portability, Reusability and Self-descriptiveness

This chapter will address four major topics (1) extensibility, (2) portability, (3) reusability, and (4) self-descriptiveness. Each of these topics are associated with adaptation concerns in design endeavors. The chapter begins by reviewing extensibility, its definitions, and how it is approached as an aspect of purposeful systems design.

The second section defines portability, provides a perspective on why portability is a desired characteristic, and four factors designers must consider in order to achieve portable designs.

The third section in this chapter addresses reusability by providing a clear definition and an addressing reusability in systems designs. Design for reuse can be achieved by using either a top-down or bottom-up approach and three unique techniques. The section concludes by recommending 2 strategies and 10 heuristics that support reusability in systems designs.

The fourth section defines self-descriptiveness and discusses the types of problems associated with poor self-descriptiveness. The section also discusses how

K.M. Adams, *Non-functional Requirements in Systems Analysis and Design*,
Topics in Safety, Risk, Reliability and Quality 28,
DOI 10.1007/978-3-319-18344-2_10

utilization of the seven design principles for user-systems dialogue and adoption and application of an appropriate standard for user-system dialogue can decrease errors and improve system self-descriptiveness.

The final section defines a measure and a means for measuring adaptation concerns that is a function of extensibility, portability, reusability, and self-descriptiveness. The section completes by relating the proposed measure for adaptation concerns as a metric and includes a structural map for extensibility, portability, reusability, and self-descriptiveness.

The chapter has a specific learning goal and associated objectives. The learning goal of this chapter is to be able to identify how the attributes of adaptability, flexibility, modifiability and scalability, and robustness that influence design in systems endeavors. This chapter's goal is supported by the following objectives:

- Define extensibility.
- Discuss how extensibility is achieved during purposeful systems design.
- Define portability.
- Describe the four factors that affect portability in systems designs.
- Define reusability.
- Describe the two approaches to reusability that may be used during design endeavors.
- Define self-descriptiveness.
- Describe the three levels of problems associated with poor self-descriptiveness.
- Construct a structural map that relate adaptation concerns to a specific metric and measureable characteristic.
- Explain the significance of extensibility, portability, reusability, and self-descriptiveness in systems design endeavors.

The ability to achieve these objectives may be fulfilled by reviewing the materials in the sections that follow.

10.2 Extensibility

In this section the basics of extensibility and how it is applied during systems endeavors will be reviewed. As with many of the other non-functional requirements, extensibility is not well understood or used in ordinary discussions about systems requirements. To validate this assertion, take a minute and review the index of a systems engineering or software engineering text and look for the word *extensibility*. Is it missing? It would not be surprising to hear that the word is missing from just about every major text. Therefore, extensibility and its characteristics must be carefully reviewed in order to provide a common base for both learning and application during systems design endeavors.

10.2.1 Definition for Extensibility

Extensibility or extendability, from a systems engineering perspective, is defined as:

> The ease with which a system or component can be modified to increase its storage or functional capacity. Syn: expandability, extensibility (IEEE and ISO/IEC 2010 p. 136).

Extensibility (which is preferred over the word extendability) has additional definitions, shown in Table 10.1 that may provide further meaning for the term when applied as a non-functional requirement for a system.

Additional meaning for extensibility may be obtained by reviewing the definition for its synonym, expandability, which is shown in Table 10.2.

From these definitions extensibility may be defined as *the ability to extend a system, while minimizing the level of effort required for implementing extensions and the impact to existing system functions*. Having settled on this basic definition, the next section will discuss how extensibility may be used as a purposeful element during systems design endeavors.

10.2.2 Extensibility in Systems Design

Extensibility considerations during systems design endeavors may be achieved by employing two broad, complementary methods: (1) product-line architectures and, (2) domain-specific languages. Some examples of each approach, for both hardware and software, will be provided.

Table 10.1 Additional definitions for extensibility

Definition	Source
"The property that simple changes to the design of a software artifact require a proportionally simple effort to modify its source code. Extensibility is a result of premeditated engineering, whereby anticipated variabilities in a domain are made simple by design"	Batory et al. (2002, p. 211)
"A system design principle that takes into consideration future growth by providing the ability to extend a system, while minimizing the level of effort required for implementing extensions and the impact to existing system functions"	Lopes et al. (2005, p. 368)
"The ability to extend a software system with new features and components without loss of functionality or qualities specified as requirements"	Henttonen et al. (2007, p. 3)

Table 10.2 Definitions for expandability (synonym of extensibility)

Definition	Source
Expandability is "the degree of effort required to improve or modify software functions' efficiency"	IEEE and ISO/IEC (2010, p. 135)

Extensibility has been practiced in the design of most hardware products for many years. For instance, imagine purchasing a car that has a design that would prohibit the addition of optional accessories. The dealers would be required to roll-the-dice when they made their selection of stock for their lots and customers would have to make trade-offs that they were not comfortable with. Instead, local dealers are able to add factory options because the design included the ability to add components. The ability to add components to a design is enabled by common interfaces and established standards. Designs that include established standards for common interfaces for component connections are able to accept new technology in a manner that permits seamless integration. One-of-a-kind designs with a lack of standardization are notoriously unable to accept new technological improvements. The electronics industry includes extensibility as a major non-functional requirement that permits components to be interconnected based on designs that routinely include interface points based on accepted industry standards.

The same can be said for most modern software products. Vendors that provide large enterprise resource planning (ERP) suites have modular designs for their products that permit consumers to select any number of individual software modules that perform specific business functions (e.g., financial accounting, human resource management, etc.). The vendor's architecture includes interfaces points between their own functional modules and other 3rd party vendors that provide unique support applications (e.g., customer resource management, scheduling, etc.). Modern software frameworks also include the ability to incorporate extensibility. For instance, Microsoft has developed the Managed Extensibility Framework (MEF) as a library within its .NET development framework for creating lightweight, extensible applications. MEF's components, called *parts*, declaratively specifies both the part's dependencies or *imports* and what capabilities or *exports* it makes available. When a programmer creates a part, the MEF composition engine satisfies its imports with what is available from other parts. Similarly, Oracle's Enterprise Manager has an Extensibility Exchange which is a library where programmers are able to find plug-ins and connectors they may utilize.

In conclusion, extensibility is a purposeful design function that permits systems to be extended—added to or modified—during their design life with a minimum of effort and subsequent disruption to the system and its users. The eminent software pioneer David Lorge Parnas makes the point about how engineers must account for change during the design stages when he states "One of the clearest morals in the earlier discussion about *design for change* as it is taught in other areas of engineering is that one must anticipate changes before one begins the design" (Parnas 1979, p. 130).

The section that follows will address the non-functional requirement for portability.

10.3 Portability

In this section the basics of portability and how it is applied during systems endeavors will be addressed. As with many of the other non-functional requirements, portability is not well understood or used in ordinary discussions about systems requirements. Once again, take a minute and review the index of a favorite systems engineering or software engineering text and look for the word portability. Is it missing? It would not be surprising to hear that the word is missing from just about every major text. Therefore, a careful review of portability and its characteristics is in order to provide a common base for both learning and its application during systems design endeavors.

10.3.1 Definition for Portability

Portability, from a systems engineering perspective, is defined as:

> The ease with which a system or component can be transferred from one hardware or software environment to another. (IEEE and ISO/IEC 2010, p. 261)

Portability has additional definitions, from the literature, that are listed in Table 10.3 that may provide further help in understanding the term when applied as a non-functional requirement for a system.

From these definitions portability is *the degree to which a system can be transported or adapted to operate in a new environment*. Having settled on this basic definition, the next section will discuss how portability may be used as a purposeful element during systems design endeavors.

10.3.2 Portability in Systems Design

In Chap. 9 it was stated that changes occur in systems based upon one or more of the following major factors: (1) marketplace forces; (2) technological evolution; and (3) environmental shifts (Fricke and Schulz 2005). Because real-world systems exist in a constantly changing domain, they too are subject to change and these

Table 10.3 Additional definitions for portability

Definition	Source
"An application is portable across a class of environments to the degree that the effort required to transport and adapt it to a new environment in the class is less than the effort of redevelopment"	Mooney (1990, p. 59)
"Effort required to transfer the program from one hardware and/or software system environment to another"	Pressman (2004, p. 510)

drivers of change must be addressed by systems designers as a practical, ever-present element of the system's lifecycle. As a result, designing a system that has contains some degree of portability that is, an ability to be transported or adapted to operate in a new environment, seems to make eminent sense.

> The primary goal of portability is to facilitate the activity of porting an application from an environment in which it currently operates to a new or target environment. This activity has two major aspects: (1) transportation—physical movement of the program's instructions and data to the new environment, and (2) adaptation—modification of the information as necessary to work satisfactorily in the new environment. (Mooney 1990, p. 59)

In order to achieve systems that are portable, designers are required to address four major factors: (1) impediment factors; (2) human factors; (3) environmental, and (4) cost factors (Hakuta and Ohminami 1997).

- *Impediment factors* include all of the technical issues that cloud, restrict, or prohibit the movement of the system from its current environment to the new or target environment. Some examples of technical issues include reusability of hardware and software, compatibility of hardware, software, and standards, interfaces between hardware and software, data structures, size, and restructuring effort, compatibility of design tools as well as development and testing environments.
- *Human factors* address the knowledge and experience of the design team and their ability to address the tasks required to transport or adapt the system to operate in the new or target environment.
- *Environmental factors* address the target environment. Specifically, the "set of elements and their relevant properties, which elements are not part of the system, but a change in any of which can cause or produce a change in the state of the system" (Ackoff and Emery 2006, p. 19). New target environments are often a significant challenge to design teams unfamiliar with the elements in the new environment.
- *Cost factors* address the aggregate the individual costs attributed to the impediment, human, and environmental factors associated with the transportation and adaptation of the existing system required for it to operate in the new or target environment.

All of these factors must be addressed by the systems designer when evaluating the decision to incorporate portability requirements as part of the purposeful design during systems endeavors. The section that follows will address system reusability.

10.4 Reusability

In this section the basics of reusability and how it is applied during systems endeavors will be reviewed. Compared to the other non-functional requirements addressed so far, reusability is a term that is used frequently during discussions about systems requirements. Despite its frequent use, we will review its formal systems vocabulary definition as well as some definitions from the literature in order to solidify a common usage for the term during systems design endeavors.

10.4.1 Definition for Reusability

Reusability, from a systems engineering perspective, is defined as:

> The degree to which an asset can be used in more than one software system, or in building other assets. (IEEE and ISO/IEC 2010, p. 307)

Reusability has additional definitions, from the literature, that are listed in Table 10.4 that may provide further help in understanding the term when applied as a non-functional requirement for a system.

From these definitions reusability is *the degree to which a system repeats the use of any part of an existing system in a new system*. Having settled on this basic definition, the next section will discuss how reusability may be used as a purposeful element during systems design endeavors.

10.4.2 Reusability as an Element of Systems Design

Reusability, from a systems engineering and design perspective, is generally viewed as a positive requirements that can directly reduce the costs associated with the design and development of a new system.

> On the other hand, enabling reuse across all activities of the development process, levels of the solution structure, and different engineering disciplines is regarded [as] an effective but challenging means for reducing costs and development time and to increase solution quality. (Stallinger et al. 2011, p. 121)

The challenges associated with reusability are based upon two important characteristics required in order to effect reuse in a design. Every designer must

Table 10.4 Additional definitions for reusability

Definition	Source
"The repeated use of any part of a [software] system: documentation, code, design, requirements, test cases, test data, and more"	Pfleeger (1998, p. 477)
"Reuse is a repetition or similarity in a design"	Hornby (2007, p. 52)
"The idea to reuse previously developed engineering artifacts in the engineering of a solution. Reuse in engineering is not limited to solution components. It pervades all engineering phases and also applies to engineering artifacts like requirements specifications, use cases, architectures, documentation, etc"	Stallinger et al. (2010, p. 308)

carefully analyze the ability of an existing systems element's ability to satisfy both (1) functionality and (2) required interfaces. As a result, designers of system elements will be required to make tradeoffs between the functionality and interface requirements in their design, and the functionality and interfaces requirements of potentially reusable systems elements. Few existing system's elements provide both identical functionality and interfaces, so tradeoffs take on additional importance when purposefully including reusability as a non-functional requirement in a system's design.

An additional design consideration must be made when invoking reusability as a non-functional requirement for a system. The design team must decide on a reusability approach: Will the reuse design be top-down (often labeled generative) or bottom-up (often termed compositional)?

> Component-oriented reuse as the major bottom-up concept is based on the idea to build a system from smaller, less complex parts by reusing or adapting existing components; top-down reuse approaches—in contrast—are more challenging, as they require a thorough understanding of the overall structure of the engineered solution (Stallinger et al. 2011, p. 121).

Reuse is viewed from two high-level perspectives—that of the producer of reusable systems elements or that of the consumer of reusable elements (Bollinger and Pfleeger 1990). Table 10.5 provides a taxonomy of consumer reuse techniques that may be used when deciding on how a consumer will use the reusable element.

Finally, reuse is an organizational issue and is not limited to the designer and design team (Lim 1998; Lynex and Layzell 1998). Organizations typically adopt one of two strategic positions shown in Table 10.6 when approaching reuse in systems endeavors.

Table 10.5 Reuse techniques for consumers of reusable elements

Reuse technique	Description
Black-box reuse	The consumer will utilize the reusable element without modification
White box reuse	The consumer will evaluate whether modification to the reusable element will require more resources that to build a new component
Clear box reuse	The consumer will modify the reusable element to fit their particular requirements

Table 10.6 Organizational reuse strategies

Reuse strategy	Description
Opportunistic	"Attempting to obtain savings through reuse, but without an overarching plan for how this will be accomplished" Fortune and Valerdi (2013, p. 306)
Strategic	"Reuse is more deliberate and process-oriented; candidate products for reuse are identified upfront, and an investment in making a product more reusable may be made" Fortune and Valerdi (2013, p. 306)

In order to successfully execute reusability in systems, organizations must adopt formal reuse processes and supporting technologies if they are to effectively implement reusability practices as part of larger systems design endeavors. There are ten heuristics that describe why certain reuse properties exist and all ten may be adopted as a means of ensuring that organizations make decisions that support reusability as an organizational strategy (Fortune and Valerdi 2013).

- Heuristic #1: Reuse is not free, upfront investment is required (i.e., there is a cost associated with design for reusability).
- Heuristic #2: Reuse needs to be planned from the conceptualization phase of programs.
- Heuristic #3: Most project related products can be reused.
- Heuristic #4: Reuse is more successful when level of service requirements are equivalent across applications.
- Heuristic #5: Reuse is as much of an organizational issue as it is a technical one.
- Heuristic #6: The benefits of reuse are nonlinear with regard to project size (i.e., small-scale systems have been shown to not enjoy the same benefit from reuse as large-scale systems).
- Heuristic #7: Higher reuse opportunities exist when there is a match between the diversity and volatility of a product line and its associated supply chain.
- Heuristic #8: Bottom-up (individual elements where make or buy decisions are made) and top-down (where product line reuse is made) reuse require fundamentally different strategies.
- Heuristic #9: Reuse applicability is often time dependent; rapidly evolving domains offer fewer reuse opportunities than static domains (i.e., products may have a "reuse" shelf life).
- Heuristic #10: The economic benefits of reuse can be described either in terms of improvement (quality, risk identification) or reduction (defects, cost/effort, time to market) (pp. 305–306).

All of these factors must be addressed by the systems designer when evaluating the decision to incorporate reusability requirements as part of a purposeful design during systems endeavors. The section that follows will address system self-descriptiveness.

10.5 Self-descriptiveness

In this section the basics of self-descriptiveness and how it is applied during systems endeavors will be addressed. Self-descriptiveness, when compared to the other non-functional requirements addressed so far, is a term that is rarely used during discussions about systems requirements. Because of its infrequent use in ordinary conversation, a review of both its formal systems vocabulary definition as well as some definitions from the literature are required. This will solidify a common meaning for the term during our discussions of its use during systems design endeavors.

10.5.1 Definition for Self-descriptiveness

Self-descriptiveness, from a systems engineering perspective, is defined as:

> 1. The degree to which a system or component contains enough information to explain its objectives and properties. 2. Software attributes that explain a function's implementation. cf. maintainability, testability, usability. (IEEE and ISO/IEC 2010, p. 322)

Self-descriptiveness has additional definitions, from the literature, that are listed in Table 10.7 that may be used to understand the term when applied as a non-functional requirement for a system.

From these definitions self-descriptiveness *is the characteristic of a system that permits an observer to determine or verify how its functions are achieved.* Having settled on this basic definition, the next section will discuss how self-descriptiveness is achieved during systems design endeavors.

10.5.2 Self-descriptiveness in Systems Design

The concept that underlies self-descriptiveness is related to the *dialogue* the user has with the system under consideration. In this context dialogue is defined as the "interaction between a user and an interactive system as a sequence of user actions (inputs) and system responses (outputs) in order to achieve a goal" (ISO 2006, p. vi). More simply stated, *dialogue is the interaction between a user and the system of interest required to achieve a desired goal.* As part of this dialogue a system designer must strive to understand how the system and its user (be that the

Table 10.7 Additional definitions for self-descriptiveness

Definition	Source
"It contains enough information for a reader to determine or verify its objectives, assumptions, constraints, inputs, outputs, components, and revision status"	Boehm et al. (1976, p. 600)
"Those characteristics [of software] which provide explanation of the implementation of functions"	Bowen et al. (1985, p. 3–12)
"Information feedback, user guidance and support"	Park and Lim (1999, p. 312)
"If every single dialogue step can immediately be understood by the user based on the information displayed by the system or if there is a mechanism to obtain any other explanatory information on request of the user"	ISO (2006, p. 6)
(For a model) "The ability of the model concepts to embed enough information to explain the model objectives and properties"	Ben Ahmed et al. (2010, p. 110)
"A dialog is self-descriptive if every single dialog step can immediately be understood by the user based on the information displayed by the system"	Frey et al. (2011, p. 268)

Table 10.8 Problems associated with self-descriptiveness

Problem	Description
Lack of self-descriptiveness	The information about the system presented to the user was missing
Perceptual problems with self-descriptiveness	The information about the system presented to the user offered some form of self-descriptiveness but was inadequate to establish understanding
Conceptual problems of self-descriptiveness	The information about the system presented to the user is clearly displayed to the user but there is some sort of cognitive barrier preventing understanding

designer or the person utilizing the system to complete its intended functions) communicate. International Standard 9241, Part 110—Dialogue Principles (ISO 2006) lists seven principles, the second of which is self-descriptiveness, which define how usable designs should behave. Research has demonstrated "that among the dialogue principles, self-descriptiveness is the most important" (Watanabe et al. 2009, p. 825).

Table 10.8 describes three levels of problems that are associated with self-descriptiveness.

By adopting the seven design principles for user-system dialogue, designers can greatly reduce (1) generalized dialogue errors, (2) the specific self-descriptiveness errors described in Table 10.8, and (3) the broader range of seven systems errors (Adams and Hester 2012, 2013). Inclusion of the non-functional requirement for self-descriptiveness in a design requires the design team to formally adopt and apply an appropriate standard for user-system dialogue (ISO 2006).

Self-descriptiveness' importance moves beyond the system design and directly impacts the system's implementation and continued viability during the operation and maintenance stage and through retirement and disposal. It is precisely because the user-system dialogue has such lasting effects that it gains importance in systems design endeavors. "Self-descriptiveness is necessary for both testability and understandability" (Boehm et al. 1976, p. 606) and can be extrapolated to other requirements as well.

The next section will discuss how the four non-functional requirements for extensibility, portability, reusability, and self-descriptiveness can be measured and evaluated.

10.6 A Method for Evaluating Extensibility, Portability, Reusability and Self-descriptiveness

The ability to understand, measure, and evaluate the non-functional requirements for extensibility, portability, reusability, and self-descriptiveness when included as requirements in a system is a valuable capability. Having the ability to measure and evaluate each of these non-functional requirements provides additional perspectives and insight into the future performance and viability of all elements of the system being designed.

Having established a basic understanding about extensibility, portability, reusability, and self-descriptiveness and how they are used in systems design endeavors, a method to measure them must be developed. Development of a satisfactory measure is problematic because each of these non-functional requirements are subjective, qualitative measures which differ greatly from most of the objective, quantitative measures developed for many of the other non-functional requirements that have been addressed. In order to understand how to approach a subjective, qualitative measure, as done in the previous chapter, a review of how to construct and measure subjective, qualitative objects is required.

10.6.1 Development of Measurement Scales

In order to evaluate extensibility, portability, reusability, and self-descriptiveness, questions that address both the presence (yes or no) and quality of the effort (how well) to provide each of the non-functional requirements as a purposeful effort during a system design endeavor must be addressed. The goal is to frame each non-functional requirements as an object with a specific measureable attribute. The establishment of effective and appropriate measures is important because they are the linkage between the observable, real-world, empirical facts about the system and the construct (i.e., extensibility, portability, reusability, and self-descriptiveness) devised to act as evaluation points. In this case, a measure is defined as "… an observed score gathered through self-report, interview, observation, or some other means" (Edwards and Bagozzi 2000, p. 156).

10.6.1.1 Scales for Extensibility, Portability, Reusability, and Self-descriptiveness

As discussed during the development of the scales for traceability (see Chap. 6), system safety (see Chap. 8), and changeability (see Chap. 9) the selection of a measurement scale is an important element in the development of an adequate measure. Because none of the non-functional requirements selected as criteria has a natural origin or empirically defined distance, an ordinal scale should be selected as an appropriate scale for measuring system extensibility, portability, reusability, and self-descriptiveness. In order to ensure improved reliability, a five-point Likert scale will be invoked (Lissitz and Green 1975).

10.6.1.2 Proposed Scales

As mentioned in previous chapters, scales are characterized as either a *proposed* scale or a scale. "A proposed scale is one that some investigator(s) put forward as having the requisite properties, and if it is indeed shown to have them, then it is recognized as a scale" (Cliff 1993, p. 65). In this chapter, the use of the word scale is referring to *proposed scales*. This may seem to be an insignificant point, but until the scale has been accepted and successfully utilized, it remains proposed.

10.6.1.3 Proposed Measurement Scale for Extensibility, Portability, Reusability, and Self-descriptiveness

Armed with a construct, measurement attributes, and an appropriate scale type, the measures for extensibility, portability, reusability, and self-descriptiveness may be constructed. In order to evaluate these, questions that address both the presence (yes or no) and quality of the effort (how well) to provide effective and meaningful levels of extensibility, portability, reusability, and self-descriptiveness as part of the system's design must be addressed. Each of the four criteria (i.e., the measurement constructs) has a specific question, shown in Table 10.9, which may be used to evaluate each one's contribution to system adaptation concerns.

The answer to each question in Table 10.9 will be scored using the five-point Likert measures in Table 10.10.

The summation of the four constructs in Eq. 10.1 will be the measure of the degree of adaptation in a system design endeavor.

Expanded Equation for System Adaptability Concerns

$$A_{sys} = A_{exten} + A_{port} + A_{reuse} + A_{selfdes} \tag{10.1}$$

Table 10.9 Measurement questions for adaptation concerns

Measurement construct	Adaptation concern for measurement
A_{extens}	Does the ability to extend the system, while minimizing the level of effort required for implementing extensions and the impact to existing system functions exist?
A_{port}	Can the system be transported or adapted to operate in a new environment?
A_{reuse}	Does the system repeat the use of any part of an existing system in its design or construction?
$A_{selfdes}$	Does the system have characteristic that permits an observer to determine or verify how its functions are achieved?

Table 10.10 Adaptation measurement question Likert scale

Measure	Descriptor	Measurement criteria
0.0	None	No objective quality evidence is present
0.5	Limited	Limited objective quality evidence is present
1.0	Nominal	Nominal objective quality evidence is present
1.5	Wide	Wide objective quality evidence is present
2.0	Extensive	Extensive objective quality evidence is present

Table 10.11 Four-level structural map for measuring adaptation concerns

Level	Role
Concern	Systems adaptation
Attribute	Adaptation concerns
Metrics	Extensibility, portability, reusability, and self-descriptiveness
Measurable characteristics	Sum of (1) extensibility (A_{exten}), (2) portability (A_{port}), (3) reusability (A_{reuse}), and (4) self-descriptiveness ($A_{selfdes}$)

10.6.2 Measuring Extensibility, Portability, Reusability and Self-descriptiveness

At the end of Chap. 3 the importance of being able to measure each non-functional attribute was emphasized. A structural mapping that relates adaptation concerns to four specific metrics and measurement entities is required. The four-level construct for adaptation concerns is presented in Table 10.11.

10.7 Summary

In this chapter, the adaptation concerns contained in the four non-functional requirements: extensibility, portability, reusability, and self-descriptiveness have been addressed. In each case, a formal definition has been provided along with additional explanatory definitions and terms. The ability to effect each of the four the non-functional requirements during the design process has also been addressed. Finally, a formal metric and measurement characteristic have been proposed for evaluating design concerns through metrics for extensibility, portability, reusability, and self-descriptiveness.

The chapter that follows will address non-functional requirements associated with system viability concerns.

References

Ackoff, R. L., & Emery, F. E. (2006). *On Purposeful Systems—An Interdisciplinary Analysis of Individual and Social Behavior as a System of Purposeful Events*. Piscataway, NJ: Aldine.

Adams, K. M., & Hester, P. T. (2012). Errors in Systems Approaches. *International Journal of System of Systems Engineering, 3*(3/4), 233–242.

Adams, K. M., & Hester, P. T. (2013). Accounting for errors when using systems approaches. *Procedia Computer Science, 20*, 318–324.

Batory, D., Johnson, C., MacDonald, B., & von Heeder, D. (2002). Achieving extensibility through product-lines and domain-specific languages: A case study. *ACM Transactions on Software Engineering Methodology, 11*(2), 191–214.

Ben Ahmed, W., Mekhilef, M., Yannou, B., & Bigand, M. (2010). Evaluation framework for the design of an engineering model. *Artificial Intelligence for Engineering Design, Analysis and Manufacturing, 24*(Special Issue 01), 107–125.

Boehm, B. W., Brown, J. R., & Lipow, M. (1976). Quantitative evaluation of software quality. In R. T. Yeh & C. V. Ramamoorthy (Eds.), *Proceedings of the 2nd International Conference on Software Engineering* (pp. 592–605). Los Alamitos, CA: IEEE Computer Society Press.

Bollinger, T. B., & Pfleeger, S. L. (1990). Economics of reuse: issues and alternatives. *Information and Software Technology, 32*(10), 643–652.

Bowen, T. P., Wigle, G. B., & Tsai, J. T. (1985). *Specification of software quality attributes: Software quality evaluation guidebook (RADC-TR-85–37, Vol III)*. Griffiss Air Force Base, NY: Rome Air Development Center.

Cliff, N. (1993). What is and isn't measurement. In G. Keren & C. Lewis (Eds.), *A Handbook for Data Analysis in the Behavioral Sciences: Methodological Issues* (pp. 59–93). Hillsdale, NJ: Lawrence Erlbaum Associates.

Edwards, J. R., & Bagozzi, R. P. (2000). On the nature and direction of relationships between constructs and measures. *Psychological Methods, 5*(2), 155–174.

Fortune, J., & Valerdi, R. (2013). A framework for reusing systems engineering products. *Systems Engineering, 16*(3), 304–312.

Frey, A. G., Céret, E., Dupuy-Chessa, S., & Calvary, G. (2011). QUIMERA: A quality metamodel to improve design rationale, *Proceedings of the 3rd ACM SIGCHI symposium on Engineering interactive computing systems* (pp. 265–270). New York: Association for Computing Machinery.

Fricke, E., & Schulz, A. P. (2005). Design for changeability (DfC): Principles to enable changes in systems throughout their entire lifecycle. *Systems Engineering, 8*(4), 342–359.

Hakuta, M., & Ohminami, M. (1997). A study of software portability evaluation. *Journal of Systems and Software, 38*(2), 145–154.

Henttonen, K., Matinlassi, M., Niemelä, E., & Kanstrén, T. (2007). Integrability and extensibility evaluation from software architectural models—a case study. *Open Software Engineering Journal, 1*, 1–20.

Hornby, G. S. (2007). Modularity, reuse, and hierarchy: Measuring complexity by measuring structure and organization. *Complexity, 13*(2), 50–61.

IEEE, & ISO/IEC (2010). IEEE and ISO/IEC Standard 24765: Systems and software engineering —vocabulary. New York and Geneva: Institute of Electrical and Electronics Engineers and the International Organization for Standardization and the International Electrotechnical Commission.

ISO. (2006). *ISO standard 9241-110: Ergonomics of human-system interaction—part 110: Dialogue principles*. Geneva: International Organization for Standardization.

Lim, W. C. (1998). Strategy-driven reuse: Bringing reuse from the engineering department to the executive boardroom. *Annals of Software Engineering, 5*(1), 85–103.

Lissitz, R. W., & Green, S. B. (1975). Effect of the number of scale points on reliability: A Monte Carlo approach. *Journal of Applied Psychology, 60*(1), 10–13.

Lopes, T. P., Neag, I. A., & Ralph, J. E. (2005). The role of extensibility in software standards for automatic test systems, *Proceedings of AUTOTESTCON—The IEEE systems readiness technology conference* (pp. 367–373). Piscataway, NJ: Institute of Electrical and Electronics Engineers.

Lynex, A., & Layzell, P. J. (1998). Organisational considerations for software reuse. *Annals of Software Engineering, 5*(1), 105–124.

Mooney, J. D. (1990). Strategies for supporting application portability. *Computer, 23*(11), 59–70.

Park, K. S., & Lim, C. H. (1999). A structured methodology for comparative evaluation of user interface designs using usability criteria and measures. *International Journal of Industrial Ergonomics, 23*(5–6), 379–389.

Parnas, D. L. (1979). Designing software for ease of extension and contraction. *IEEE Transactions on Software Engineering, SE-5*(2), 128–138.

Pfleeger, S. L. (1998). *Software engineering: Theory and practice*. Upper Saddle River, NJ: Prentice-Hall.

Pressman, R. S. (2004). *Software engineering: A practitioner's approach* (5th ed.). New York: McGraw-Hill.

Stallinger, F., Neumann, R., Vollmar, J., & Plösch, R. (2011). Reuse and product-orientation as key elements for systems engineering: aligning a reference model for the industrial solutions business with ISO/IEC 15288, *Proceedings of the 2011 International Conference on Software and Systems Process* (pp. 120–128). New York: ACM.

Stallinger, F., Plösch, R., Pomberger, G., & Vollmar, J. (2010). Integrating ISO/IEC 15504 conformant process assessment and organizational reuse enhancement. *Journal of Software Maintenance and Evolution: Research and Practice, 22*(4), 307–324.

Watanabe, M., Yonemura, S., & Asano, Y. (2009). Investigation of web usability based on the dialogue principles. In M. Kurosu (Ed.), *Human Centered Design* (pp. 825–832). Berlin: Springer.

Part V
Viability Concerns

Chapter 11
Understandability, Usability, Robustness and Survivability

Abstract The design of systems and components during the design stage of the systems life cycle requires specific purposeful actions to ensure effective designs and viable systems. Designers are faced with a number of *core viability concerns* that they must embed into the design to ensure the system remains viable. The ability for a system to remain viable is critical if it is to continue to provide required functionality for its stakeholders. Core viability concerns includes the non-functional requirements for understandability, usability, robustness, and survivability. Purposeful design requires an understanding of each of these requirements and how to measure and evaluate each as part of an integrated systems design.

11.1 Introduction to Understandability, Usability, Robustness and Survivability

This chapter will address four major topics (1) understandability, (2) usability, (3) robustness, and (4) survivability. Each of these topics are associated with core viability concerns in design endeavors. The chapter begins by reviewing understandability, its definitions, and how it is approached as an aspect of purposeful systems design.

Section 11.3 defines usability, provides a perspective on why usability is a desired characteristic, and describes the four attributes traditionally associated with usability.

Section 11.4 in this chapter addresses robustness by providing a clear definition and addressing robustness as an element of systems designs. Design and the concepts associated with robustness conclude the section.

Section 11.5 defines survivability and its three major contributing elements. The section also discusses 17 design principles that may be invoked when designing for survivability.

Section 11.6 defines a measure and a means for measuring core viability concerns that is a function of understandability, usability, robustness, and survivability.

K.M. Adams, *Non-functional Requirements in Systems Analysis and Design*,
Topics in Safety, Risk, Reliability and Quality 28,
DOI 10.1007/978-3-319-18344-2_11

The section completes by relating the proposed measure for core viability concerns as a metric and includes a structural map for understandability, usability, robustness, and survivability.

The chapter has a specific learning goal and associated objectives. The learning goal of this chapter is to be able to identify how the attributes of adaptability, flexibility, modifiability and scalability, and robustness that influence design in systems endeavors. This chapter's goal is supported by the following objectives:

- Define understandability.
- Discuss how understandability is achieved during purposeful systems design.
- Define usability.
- Describe the four attributes traditionally associated with usability.
- Define robustness.
- Discuss the element associated with designing for robustness.
- Define survivability.
- Describe the three elements of survivability.
- Discuss some of the design principles associated with survivability.
- Construct a structural map that relate core viability concerns to a specific metric and measurable characteristic.
- Explain the significance of understandability, usability, robustness, and survivability in systems design endeavors.

The ability to achieve these objectives may be fulfilled by reviewing the materials in the sections that follow.

11.2 Understandability

In this section the basics of understandability and how it is applied during systems endeavors will be reviewed. The definition for the non-functional requirement for understandability would seem to be a clear cut one and a topic of discussion when developing systems. However, when a search of the many texts and scholarly articles on design is conducted, little mention of understandability is revealed. In order to improve upon this situation, and to be precise and to develop an appropriate measure for usability, a common definition and associated terminology for understandability must be constructed in order to describe its use in systems design endeavors.

11.2.1 Definition for Understandability

Understandability, from a systems engineering perspective, is defined as:

> The ease with which a system can be comprehended at both the system-organizational and detailed-statement levels. NOTE Understandability has to do with the system's coherence at a more general level than readability does. (IEEE and ISO/IEC 2010, p. 385)

Table 11.1 Additional definitions for understandability

Definition	Source
"Code possesses the characteristic understandability to the extent that its purpose is clear to the inspector. This implies that variable names or symbols are used consistently, modules of code are self-descriptive, and the control structure is simple or in accordance with a prescribed standard, etc"	Boehm et al. (1976, p. 605)
"Ease of understanding the function of a program and its functional implementation."	Blundell et al. (1997, p. 237)
"It is related to the ease of use, respectively the effort for reading and correctly interpreting a conceptual model, which is a cognitive process of assigning meaning to the different parts of a conceptual model"	Houy et al. (2012, p. 66)
"The requirements are correctly understood without difficulty"	Génova et al. (2013, p. 27)

Table 11.2 Definitions for unambiguity

Definition	Source
"Described in terms that only allow a single interpretation, aided, if necessary, by a definition"	IEEE and ISO/IEC (2010, p. 384)
"There exists only one interpretation for each requirement"	Génova et al. (2013, p. 27)

Understandability has additional definitions, shown in Table 11.1 that may provide further meaning for the term when applied as a non-functional requirement for a system.

Additional meaning for understandability may be obtained by reviewing the definition for a closely related word, unambiguity. It is important to note that "unambiguity and understandability are interrelated (according to some, they would be even the same property), since, if a requirement is ambiguous, then it cannot be properly understood" (Génova et al. 2013, p. 28). Two formal definitions for unambiguity are presented in Table 11.2.

From these definitions, from a systems user's perspective, understandability is *the ability to comprehend any portion of a system without difficulty*. Having settled on this basic definition, the next section will discuss the elements that contribute to human understandability in systems.

11.2.2 Elements of Understandability

Empirical research has demonstrated that understandability in systems is the sum of three components: (1) domain knowledge, (2) contribution from the system being considered, and (3) chance. When understandability is evaluated using the sum of the three measures it is referred to as "absolute understandability" and when the measure of understandability includes only the contribution made by the system and its associated design artifacts, it is referred to as "relative understandability" (Ottensooser et al. 2012, p. 600).

Knowing the three components that contribute to systems understandability, and that *relative understandability* is a function of the system's designers and their ability to create comprehendible design artifacts, the next section will discuss how understandability may be used as a purposeful element during systems design endeavors. Each time the term understandability is mentioned in the remainder of this chapter, it is referring to *relative understandability*, over which the system's designers have ultimate control.

11.2.3 Understandability in Systems Design

Knowing that the system's designers are directly responsible for the understandability of the system and that this is a function of readily comprehendible design artifacts (e.g., user documentation, help screens, the functional flow of information; application of business rules, etc.), the design must adopt formal approaches to ensure understandability. Failure to adopt such an approach may lead to the following scenario for both users and the system's maintainers.

> If we can't learn something, we won't understand it. If we can't understand something, we can't use it—at least not well enough to avoid creating a money pit. We can't maintain a system that we don't understand—at least not easily. And we can't make changes to our system if we can't understand how the system as a whole will work once the changes are made. (Nazir and Khan 2012, p. 773)

In order to avoid this type of pitfall, systems designs must adopt and implement a clear conceptual model for understandability. "To me, the most important part of a successful design is the underlying conceptual model. This is the hard part of design: formulating an appropriate conceptual model and then assuring that everything else be consistent with it" (Norman 1999, p. 39).

In simple terms a conceptual model of a system is a mental model that people use to represent their individual understanding of how the system works. The conceptual or mental *model* is just that, it is a *perceived* structure for the real-world system that they are encountering. The utility of the conceptual model includes both (1) prediction—how things will behave and (2) functional understanding—relationships between the system's components and the functions they perform. A conceptual model "specifies both the static and the dynamic aspects of the application domain. The static aspect describes the real world entities and their relationships and attributes. The dynamic aspect is modeled by processes and their interfaces and behavior" (Kung 1989, p. 1177). The static portion of conceptual models include things and their associated properties while the dynamic aspect addresses events and their supporting processes. The static and dynamic perspectives of conceptual models have four purposes (Kung and Solvberg 1986):

1. supporting communication between systems designers and users,
2. helping analysts understand a domain,
3. providing input for the design process, and
4. documenting the original requirements for future reference.

A principle that is particularly useful in the construction of conceptual models is *The Metaphor Principle*, which states:

> People develop new cognitive structures by using metaphors to cognitive structures they have already learned. (Carroll and Thomas 1982, p. 109)

A metaphor is defined as "a figure of speech (or a trope) in which a word of phrase that literally denote one thing is used to denote another, thereby implicitly comparing the two things" (Audi 1999, p. 562). The metaphor is a powerful aid in the development of conceptual models. *The Metaphor Principle* underlies the use of metaphors as a powerful technique for accessing and using existing knowledge to create a relationship that may be used to develop new knowledge structures. "To the extent that a system can be explained to new users in terms of things that the user is already familiar with, the system will be more readily comprehended" (Branscomb and Thomas 1984, p. 233).

The development of a conceptual model is a design technique that allows the system's designers to clarify and consistently express the meaning of terms and associated concepts for the system during the design processes and the equally important operations and maintenance stages of the systems life cycle. "Completeness, consistency, understandability and unambiguity depend on precision: a more precise language helps to write requirements that are more complete, consistent, understandable and unambiguous" (Génova et al. 2013, p. 28).

Figure 11.1 shows the interrelationship of the designer's and users' conceptual models and the real-world *system image*. The system image is how the system's designers communicate with the users of the system.

The designer's conceptual model of the system, plus documentation, instructions, web-sites, help-screens, and signifiers (i.e., an indicator of system behavior provided to a user) is what constitutes the system image. The system image is the physical structure and information about the system from which the users construct their own conceptual model of the system. Deviations between the designer's conceptual model and the user's conceptual model are design faults and not only reduce understandability, but ultimately, the next topic, system usability.

In conclusion, understandability is a purposeful design function that permits systems to be comprehended by their users, with a minimum of effort, throughout the entire systems' life cycle. Norman, author of *The Design of Everyday Things* (2013) and expert on human information processing states "The most important design tool is that of coherence and understandability, which comes through an explicit, perceivable conceptual model" (Norman 1999, p. 41).

The section that follows will address the non-functional requirement for usability.

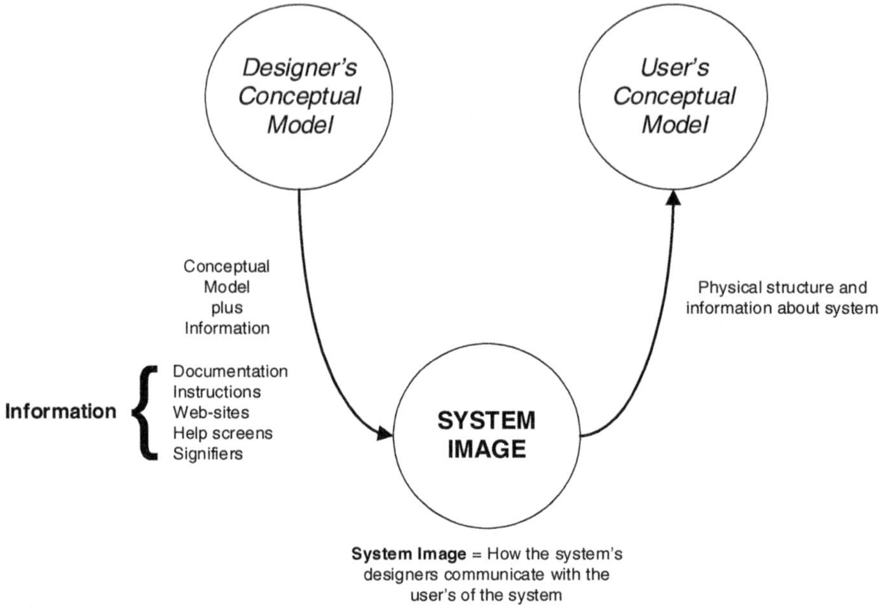

Fig. 11.1 System image and the user's and designer's conceptual models

11.3 Usability

In this section, the basics of usability and how it is applied during systems endeavors will be addressed. As with many of the other non-functional requirements, usability sounds like a very clear concept. However, arriving at a common definition and understanding for the term is problematic. Once again, take a minute and review the index of a favorite systems engineering or software engineering text and look for the word usability. Is it missing? It would not be surprising to hear that the word is missing from just about every major text. Therefore, a careful review of usability and its characteristics is in order to provide a common base for both learning and its application during systems design endeavors.

11.3.1 Definition for Usability

Usability, from a systems engineering perspective, is defined as:

> The ease with which a user can learn to operate, prepare inputs for, and interpret outputs of a system or component. (IEEE and ISO/IEC 2010, p. 388)

Usability has additional definitions, from the literature, that are listed in Table 11.3 that may provide further help in understanding the term when applied as a non-functional requirement for a system.

Table 11.3 Additional definitions for usability

Definition	Source
"Effort required to learn, operate, prepare input, and interpret output of a program"	Cavano and McCall (1978, p. 136)
"The rating is in terms of effort required to improve human factors to acceptable level"	McCall and Matsumoto (1980, p. 28)
"How easy is it to use." "For example, operability arid training are criteria for usability"	Bowen et al. (1985, pp. 2-17 and 3-1)
"A set of attributes that bear on the effort needed for use, and on the individual assessment of such use, by a stated or implied set of users"	Bevan (2001, p. 537); ISO/IEC (1991)
"Effort to learn, operate, prepare input and interpret output"	Blundell et al. (1997, p. 237)
"Degree to which a product or system can be used by specified users to achieve specified goals with effectiveness, efficiency and satisfaction in a specified context of use"	ISO/IEC (2011, 4.2.4)
"The characteristic usability subsumes the aspects of how easy the system can be used. This includes how fast using it can be learned but also if the interface is attractive and if challenged persons are able to use it"	Wagner (2013, p. 62)

From these definitions, usability is *the degree of effort required to learn, interpret, and effectively and efficiently operate a system.* Having settled on this basic definition, the next section will discuss how usability may be used as a purposeful element during systems design endeavors.

11.3.2 Usability in Systems Design

As presented in the previous section, the *system image* is the physical structure and information about the system from which the users construct their own conceptual model of the system. When deviations exist between the intentions represented in the designer's conceptual model and the actual understanding encapsulated in the user's conceptual model (which is a direct result of the system image), understandability and system usability are diminished.

When usability in a system's design is discussed, it is including design parameters such as ease of use, ease of learning, error protection, error recovery, and efficiency of performance (Maxwell 2001). Jakob Nielsen, one of the pioneers of human centered design, points out the importance of usability and that "usability has multiple components and is traditionally associated with these five attributes:

- *Learnability*: The system should be easy to learn so that the user can rapidly start getting some work done with the system.
- *Efficiency*: The systems should be efficient to use, so that once the learner has learned the system, a high level of productivity is possible.

- *Memorability*: The system should be easy to remember, so that the casual user is able to return to the system after some period of not using it, without having to learn everything all over again.
- *Errors*: The system should have a low error rate, so that users make few errors during the use of the system, and so that if they do make errors they can easily recover from them. Further, catastrophic errors must not occur.
- *Satisfaction*: The system should be pleasant to use, so that users are subjectively satisfied with using it; they like it." (Nielsen 1993, p. 26)

During the engineering design stages of the systems life cycle, specific processes that address usability elements are invoked. The processes are labeled *usability engineering* or *human centered design* and provide formal methods for ensuring that usability characteristics are specified early in the design stages and that they continue to be measured throughout all of the subsequent life cycle stages. Table 11.4 is a listing of some of the concerns during human-computer-interactions (HCI) and associated sample measures.

As with understandability, usability involves a strong cognitive component because users need to comprehend the system to some extent in order to utilize it. The relationship between data and information is an important element of human cognition.

> Most data is of limited value until it is processed into a useable form. Processing data into a useable form requires human intervention, most often accomplished with the use of an information system. The output of this process is information. Information is contained in descriptions, answers to questions that begin with such words as who, what, where, when, and how many. These functional operations are performed on data and transform it into information. (Hester and Adams 2014, p. 161)

During the data-information processing effort, the data must be presented in a usable form (i.e., represented) before it becomes meaningful information. The choice of representation directly affects the usability of the information. For instance, "diagrams are famously better than text for some problems, worse for others. One school of thought maintains that the difference lies in the cognitive processes of locating and indexing the components of information" (Green and Petre 1996, p. 134).

Designers must also be keenly aware of both the system domain and the user's functional requirements within that domain. Knowledge of the functional requirements and user domain assists the designer in selecting the attributes the system should include in support of positive usability. "Usable products can be designed by incorporating product features and attributes known to benefit users in particular contexts of use" (Bevan 2001, p. 542).

In summary, systems designers have a number of methods and techniques at their disposal that support the engineering of usable systems designs.

Table 11.4 Concerns during HCI [adapted from (Zhang et al. 2005, p. 522)]

HCI concern	Attribute area	Description	Sample measure items
Physical	Ergonomic	System fits our physical strengths and limitations and does not cause harm to our health	• Legible • Audible • Safe to use
Cognitive	Usability	System fits our cognitive strengths and limitations and functions as the cognitive extension of our brain	• Fewer errors and easy recovery • Easy to use • Easy to remember how to use • Easy to learn
Affective, emotional, and intrinsically motivational	Satisfaction	System satisfies our aesthetic and emotional needs, and is attractive for its own sake	• Aesthetically pleasing • Engaging • Trustworthy • Satisfying and enjoyable • Entertaining and fun
Extrinsically motivational	Usefulness	Using the system would provide rewording consequences	• Support individual's tasks • Can do some tasks that would not so without the system • Extend one's capability • Rewarding

> We have many tactics to follow to help people understand how to use our designs. It is important to be clear about the distinctions among them, for they have very different functions and implications. Sloppy thinking about the concepts and tactics often leads to sloppiness in design. And sloppiness in design translates into confusion for users. (Norman 1999, p. 41)

All of these factors discussed in this section must be addressed by the systems designer when evaluating the decision to incorporate usability requirements as part of the purposeful design during systems endeavors. The section that follows will address system robustness.

11.4 Robustness

In this section, robustness and how it may be applied during systems endeavors will be reviewed. As with many of the other non-functional requirements addressed so far, robustness is a term that is infrequently used during discussions about systems requirements. Based upon its infrequent use, a thorough review of both the formal definition from the systems vocabulary as well as some definitions from the literature is in order to solidify a common usage for the term in the sections that follow.

11.4.1 Definition for Robustness

Robustness, from a systems engineering perspective, is defined as:

> The degree to which a system or component can function correctly in the presence of invalid inputs or stressful environmental conditions cf. error tolerance, fault tolerance. (IEEE and ISO/IEC 2010, p. 313)

Robustness has additional definitions, from the literature, that are listed in Table 11.5 that may provide further help in understanding the term when applied as a non-functional requirement for a system.

From these definitions, robustness is *the ability of a system to maintain a desired characteristic despite fluctuations caused by either internal changes or its environment*. Armed with this composite definition, robustness and how it may be used as a purposeful element during systems design endeavors may be reviewed.

11.4.2 Robustness as an Element of Systems Design

Robustness, from an engineering perspective, is generally viewed as an operational characteristic that must be achieved through a purposeful design in order to permit the system to remain viable in real-world environments. Operation in real-world environments requires the system to absorb perturbations caused by both its intended users and the larger environment of which it is a part.

Table 11.5 Additional definitions for robustness

Definition	Source
"Insensitivity to variation."	Box and Fung (1993, p. 503)
"The maintenance of some desired system characteristic despite fluctuations in the behavior of its component parts or its environment"	Carlson and Doyle (2002, p. 2539)
"Characterizes a systems ability to be insensitive towards changing environments. Robust systems deliver their intended functionality under varying operating conditions without being changed"	Fricke and Schulz (2005, p. 347)
"The state that is insensitive to variation"	Hwang and Park (2005, p. 231)
"The ability to remain "constant" in parameters in spite of system internal and external changes"	Ross et al. (2008, p. 249)

In order to adequately ensure viability, the design process must account for both the system's users and the environments in which it will operate. It is important to note that both users and environments are plural. This is because systems must be designed to accommodate not only the currently required user base and defined environment, but changes that will add new users and which will then position the system within an ever-changing environment. Failure to do so will restrict the system's ability to operate and require expensive modification or changes to its structure. The range of operating conditions specified for the system should be broad enough to permit changes in users and environment without a significant loss of system functionality. "A product with a wide range of permissible operating conditions is more robust than a product that is more restrictive" (Schach 2002, p. 148).

11.4.2.1 Robustness Concepts

The principle method used to achieve robustness in systems has been redundancy. Redundancy "is key to flexibility and robustness since it enables capacity, functionality, and performance options as well as fault-tolerance" (Fricke and Schulz 2005, p. 355). Traditional design methods ensure that a system's elements (i.e., the components of which it is constituted) have sufficient redundancy to permit operations when some number of systems elements fail or malfunction. This is the traditional role of reliability during systems design and includes a purposeful configuration and interconnection of systems elements.

> Because robustness is achieved by very specific internal structures, when any of these systems is disassembled, there is very little latitude in reassembly if a working system is expected. Although large variations or even failures in components can be tolerated if they are designed for through redundancy and feedback regulation, what is rarely tolerated, because it is rarely a design requirement is nontrivial rearrangements of the interconnection of internal parts. (Carlson and Doyle 2002, p. 2539)

During system design endeavors, the system's designers purposefully engineer subsystems and lower level components in the system's hierarchy based on functional and non-functional requirements in the design specifications. The design is engineered module by module and the overall system reliability and robustness is a function of the system's architectural design. However, modular design and construction often weakens robustness due to the emphasis on reduced coupling in systems designs.

> The robustness of the system is embedded in its modular design; each module employs robust design principles and together the overall system maintains robustness. Robustness obtained through redundancy may suffer through modularization due to the reduction of interconnections inherent in modular design. (Ross et al. 2008, p. 258)

11.4.2.2 Designing for Robustness

Clausing (2004) advocates the use of an operating window (OW) to define the range within which a specific system characteristic is expected operate successfully.

> The OW is the range in at least one critical functional variable of the system within which the failure rate is less than some selected value. The range is bounded by thresholds (or a threshold) beyond which the performance is degraded to some selected bad level. (Clausing 2004, p. 26)

The operating window (OW) is determined by perturbing the system through introduction of a stress factor (typically termed noise) during operation. The stress factor is used to degrade the performance of the system. In the earliest stages of the system's design, the initial configuration of the system is challenged by a stress test. By challenging the system's operation early in the design process, the design team is able to verify initial failure rates and establish an operating window. The operating window is normally set between failure rates of 0.1 and 0.5 (Clausing 2004). As the system's design progresses and moves through the later stages of the design process and into the early development stage, the system is once again challenged by the stress test. The achievement of robustness may be evaluated by reviewing the amount of expansion (or contraction) of the OW that resulted from efforts to improve robustness during the final design and early development stages.

There is no one universally accepted method for assessing robustness in systems design endeavors. Robustness assessment (Huang and Du 2007) and feasibility robustness (Du and Chen 2000) are two methods that show promise when evaluating robustness.

In summary, "the desire for robustness stems from the fact that change is inevitable, both in reality and perception" (Ross et al. 2008, p. 247). Robustness is a non-functional requirement that permits systems to remain viable in the presence of both disturbances from within the system itself as well as those from its environment. Each of the factors mentioned in this section must be addressed by the systems designer when evaluating the decision to ensure robustness as part of a purposeful design during systems endeavors. The section that follows will address system survivability.

11.5 Survivability

In this section the basics of system survivability and how it is applied during systems endeavors will be addressed. Survivability, as a non-functional requirement, is a term that is very rarely used during discussions about systems requirements. Because of its infrequent use in ordinary conversation, a review the formal systems vocabulary definition and some definitions from the literature is required. This will solidify a common meaning for the term during discussions of its use and application during systems design endeavors.

11.5.1 *Definition for Survivability*

Survivability, from a systems engineering perspective, is defined as:

> The degree to which a product or system continues to fulfill its mission by providing essential services in a timely manner in spite of the presence of attacks. cf. recoverability. [SE VOCAB]

Survivability's additional definitions are taken from the literature and are listed in Table 11.6. All of these definitions may be used to better understand the term when applied as a non-functional requirement for a system.

From these definitions survivability is *the ability of a system to continue to operate in the face of attacks or accidental failures or errors.* Having settled on this basic definition, the next section will discuss how survivability is achieved during systems design endeavors.

11.5.2 *Survivability Concepts*

Generally, systems must remain survivable at all times. However, there are specific times when system survivability requirements may be relaxed. This includes times when the system is removed from operation (e.g., for upgrade, repair, or overhaul) or is placed in a reduced state of readiness (e.g., standby mode or training). The systems stakeholders should be queried about these types of situations as part of the requirements analysis process. Knowledge of limited periods during which the system does not have to remain survivable are important elements of the conceptual design for the system.

During the conceptual design of the system a number of additional items must be considered as part of the survivability approach. "Survivability requires more than simply a technical solution, but rather an integrated collaboration of technical aspects, business considerations and analysis techniques" (Redman et al. 2005, p. 187).

Table 11.6 Additional definitions for survivability

Definition	Source
"The capability of a system to fulfill its mission, in a timely manner, in the presence of attacks, failures, or accidents"	Ellison et al. (1997, p. 2)
"The capability of a system to fulfill its mission, in a timely manner and in the presence of attacks and failures"	Redman et al. (2005, p. 184)
"The ability of a system to tolerate intentional attacks or accidental failures or errors"	Korczaka and Levitin (2007, p. 269)
"The ability of a system to minimize the impact of a finite-duration disturbance on value delivery"	Mekdeci et al. (2011, p. 565)

The purposeful design of highly survivable systems requires specialized knowledge. Individuals who practice survivability engineering are focused on minimizing the impact of external disturbances (i.e. from the environment) on the system's ability to operate satisfactorily in the presence of attacks or accidental failures or errors. Conceptually, designers may achieve survivability by addressing one of three types of event (Westrum 2006):

- *Type I Event—Reducing Susceptibility*: reducing the probability that an actual disturbance may impact the system.
- *Type II Event—Reducing Vulnerability*: reducing the amount of value lost by the system or its stakeholders as a direct result of an actual disturbance.
- *Type III Event—Improving Resilience or Recoverability*: increasing the ability to return the system to service in a timely manner after an actual disturbance.

The section that follows will discuss how specific design principles may be used to enhance system survivability by addressing the three survivability event elements: (1) susceptibility, (2) vulnerability, and (3) resilience.

11.5.3 Survivability in Systems Design

In order to successfully incorporate considerations that will address susceptibility, vulnerability, and resilience during a systems design a review of how each survivability element may be addressed as an element of a purposeful design is required. There are several design approaches that propose frameworks for addressing system survivability that include both (1) the integrated engineering framework (Ellison et al. 1997) and (2) the bio-networking architecture (Nakano and Suda 2007). However, the most impressive work has been done by Mathew Richards and his colleagues at the Massachusetts Institute of Technology (Richards et al. 2008, 2009) in which they focus on the development of design principles that address each of the three survivability elements. Table 11.7 displays the three survivability elements and the associated design principles that may be invoked when addressing survivability as a purposeful activity during system design endeavors.

By focusing on the three elements of survivability and their associated design principles, the design team has an improved ability to address systems concepts based on foundational principles. By basing understanding on established design principles the designers are able to expand their range of alternatives throughout the design process.

In summary, "Survivability is particularly concerned with the successful delivery of system essential services, to ultimately fulfill mission objectives" (Redman et al. 2005, p. 185). As a non-functional requirement, survivability addresses the requirement to provide the capability to continue systems operations in the face of attacks or accidental failures or errors. The next section will discuss how the non-functional requirements for understandability, usability, robustness, and survivability may be measured and evaluated.

Table 11.7 Survivability elements and associated design principles

Survivability element	Design principle	Design principle definition
Susceptibility	Prevention	Suppression of a future or potential future disturbance
	Mobility	Relocation to avoid detection by an external change agent
	Concealment	Reduction of the visibility of a system from an external change agent
	Deterrence	Dissuasion of a rational external change agent from committing a disturbance
	Preemption	Suppression of an imminent disturbance
	Avoidance	Maneuverability away from an ongoing disturbance
Vulnerability	Hardness	Resistance of a system to deformation
	Redundancy	Duplication of critical system functions to increase reliability
	Margin	Allowance of extra capability for maintaining value delivery despite losses
	Heterogeneity	Variation in system elements to mitigate homogeneous disturbances
	Distribution	Separation of critical system elements to mitigate local disturbances
	Failure mode reduction	Elimination of system hazards through intrinsic design: substitution, simplification, decoupling, and reduction of hazardous materials
	Fail-safe	Prevention or delay of degradation via physics of incipient failure
	Evolution	Alteration of system elements to reduce disturbance effectiveness
	Containment	Isolation or minimization of the propagation of failure
Resilience	Replacement	Substitution of system elements to improve value delivery
	Repair	Restoration of system to improve value delivery

11.6 A Method for Evaluating Understandability, Usability, Robustness and Survivability

The ability to understand, measure, and evaluate the non-functional requirements for understandability, usability, robustness, and survivability when included as requirements in a system is a valuable capability. Having the ability to measure and evaluate each of these non-functional requirements provides additional perspectives and insight into the future performance and viability of all elements of the system being designed.

Based upon the understanding developed in the previous sections on understandability, usability, robustness, and survivability and how they are used in systems design endeavors, a technique for measure may be developed. Once again, this is a tough assignment because each of these non-functional requirements are subjective, qualitative measures which differ greatly from most of the objective, quantitative measures have developed for many of the other non-functional requirements. In order to understand how to approach a subjective, qualitative measure, a review of how to construct and measure subjective, qualitative objects will be presented.

11.6.1 Development of Measurement Scales

As with the other qualitative non-functional requirements, in order to satisfactorily evaluate understandability, usability, robustness, and survivability, questions that address both the presence (yes or no) and quality of the effort (how well) to provide each of the non-functional requirements during system design endeavors will need to be addressed. To support this goal objects with a specific measureable attribute will be developed. The establishment of the measures is important because they are the link between what is observed in the real-world, and as such represent the empirical facts about the system and the constructs for understandability, usability, robustness, and survivability devised as evaluation points.

11.6.1.1 Scales for Understandability, Usability, Robustness and Survivability

As discussed during the development of the scales for traceability (see Chap. 6), system safety (see Chap. 8), changeability (see Chap. 9), and adaptation concerns (see Chap. 10), the selection of a measurement scale is an important element in the development of an adequate measure. Because none of the non-functional requirements selected as criteria has a natural origin or empirically defined distance, an ordinal scale should be selected as an appropriate scale for measuring system extensibility, portability, reusability, and self-descriptiveness. In order to ensure improved reliability, a five-point Likert scale will be invoked (Lissitz and Green 1975).

11.6.1.2 Proposed Scales

As mentioned in previous chapters, scales are characterized as either a *proposed* scale or a scale. "A proposed scale is one that some investigator(s) put forward as having the requisite properties, and if it is indeed shown to have them, then it is recognized as a scale" (Cliff 1993, p. 65). In this chapter, the use of the word scale is referring to *proposed scales*.

11.6.1.3 Proposed Measurement Scale for Understandability, Usability, Robustness and Survivability

Armed with a construct, measurement attributes, and an appropriate scale type, the measures for understandability, usability, robustness, and survivability are

Table 11.8 Measurement questions for viability

Measurement construct	Adaptation concern for measurement
V_{under}	Can a person comprehend any portion of a system without difficulty?
V_{use}	What is the degree of effort required to learn, interpret, and effectively and efficiently operate a system?
V_{robust}	Does the system demonstrate the ability to maintain a desired characteristic despite fluctuations caused by either internal changes or its environment?
V_{surviv}	Does the system demonstrate the ability to continue to operate in the face of attacks or accidental failures or errors?

Table 11.9 Viability measurement question Likert scale

Measure	Descriptor	Measurement criteria
0.0	None	No objective quality evidence is present
0.5	Limited	Limited objective quality evidence is present
1.0	Nominal	Nominal objective quality evidence is present
1.5	Wide	Wide objective quality evidence is present
2.0	Extensive	Extensive objective quality evidence is present

constructed. In order to evaluate these, questions that address both the presence (yes or no) and quality of the effort (how well) to provide effective and meaningful levels of understandability, usability, robustness, and survivability as an element of the system's design must be answered. Each of the four criteria (i.e., the measurement constructs) has a specific question, shown in Table 11.8, which may be used to evaluate each one's contribution to system adaptation concerns.

The answer to each question in Table 11.8 will be scored using the five-point Likert measures in Table 11.9.

The summation of the four constructs in Eq. 11.1 will be the measure of the degree of core viability in a system design endeavor.

Expanded Equation for System Viablity Concerns

$$V_{core} = V_{under} + V_{use} + V_{robust} + V_{surviv} \tag{11.1}$$

11.6.2 Measuring Understandability, Usability, Robustness and Survivability

At the end of Chap. 3, the importance of being able to measure each non-functional attribute was stressed as an important feature. A structural mapping that relates core viability concerns to four specific metrics and measurement entities is required. The four-level construct for core viability concerns is presented in Table 11.10.

Table 11.10 Four-level structural map for measuring core viability concerns

Level	Role
Concern	Systems viability
Attribute	Core viability concerns
Metrics	Understandability, usability, robustness and survivability
Measurable characteristics	Sum of (1) understandability (V_{under}), (2) usability (V_{use}), (3) robustness (V_{robust}), and (4) survivability (V_{surviv})

11.7 Summary

In this chapter, the core adaptation concerns contained in the four non-functional requirements: understandability, usability, robustness, and survivability have been addressed. In each case, a formal definition has been provided along with additional explanatory definitions and terms. The ability to effect each of the four the non-functional requirements during the design process has also been addressed. Finally, a formal metric and measurement characteristic have been proposed for evaluating design concerns through metrics for understandability, usability, robustness, and survivability.

The chapter that follows will address additional non-functional requirements associated with system viability.

References

Audi, R. (Ed.). (1999). *Cambridge dictionary of philosophy* (2nd ed.). New York: Cambridge University Press.

Bevan, N. (2001). International standards for HCI and usability. *International Journal of Human-Computer Studies, 55*(4), 533–552.

Blundell, J. K., Hines, M. L., & Stach, J. (1997). The measurement of software design quality. *Annals of Software Engineering, 4*(1), 235–255.

Boehm, B. W., Brown, J. R., & Lipow, M. (1976). Quantitative evaluation of software quality. In R. T. Yeh & C. V. Ramamoorthy (Eds.), *Proceedings of the 2nd International Conference on Software Engineering* (pp. 592–605). Los Alamitos, CA: IEEE Computer Society Press.

Bowen, T. P., Wigle, G. B., & Tsai, J. T. (1985). *Specification of software quality attributes: Software quality evaluation guidebook* (RADC-TR-85-37, Vol. III). Griffiss Air Force Base, NY: Rome Air Development Center.

Box, G. E. P., & Fung, C. A. (1993). Quality quandries: Is your robust design procedure robust? *Quality Engineering, 6*(3), 503–514.

Branscomb, L. M., & Thomas, J. C. (1984). Ease of use: A system design challenge. *IBM Systems Journal, 23*(3), 224–235.

Carlson, J. M., & Doyle, J. (2002). Complexity and robustness. *Proceedings of the National Academy of Sciences of the United States of America, 99*(3), 2538–2545.

Carroll, J. M., & Thomas, J. C. (1982). Metaphor and the cognitive representation of computing systems. *IEEE Transactions on Systems, Man and Cybernetics, 12*(2), 107–116.

Cavano, J. P., & McCall, J. A. (1978). A framework for the measurement of software quality. *SIGSOFT Software Engineering Notes, 3*(5), 133–139.

Clausing, D. P. (2004). Operating window: An engineering measure for robustness. *Technometrics, 46*(1), 25–29.

Cliff, N. (1993). What is and isn't measurement. In G. Keren & C. Lewis (Eds.), *A Handbook for Data Analysis in the Behavioral Sciences: Methodological Issues* (pp. 59–93). Hillsdale, NJ: Lawrence Erlbaum Associates.

Du, X., & Chen, W. (2000). Towards a better understanding of modeling feasibility robustness in engineering design. *Journal of Mechanical Design, 122*(4), 385–394.

Ellison, R., Fisher, D., Linger, R., Lipson, H., Longstaff, T., & Mead, N. (1997). *Survivable network systems: An emerging discipline (CMU/SEI-97-TR-013)*. Pittsburgh: Carnegie Mellon University.

Fricke, E., & Schulz, A. P. (2005). Design for changeability (DfC): Principles to enable changes in systems throughout their entire lifecycle. *Systems Engineering, 8*(4), 342–359.

Génova, G., Fuentes, J., Llorens, J., Hurtado, O., & Moreno, V. (2013). A framework to measure and improve the quality of textual requirements. *Requirements Engineering, 18*(1), 25–41.

Green, T. R. G., & Petre, M. (1996). Usability analysis of visual programming environments: A 'cognitive dimensions' framework. *Journal of Visual Languages & Computing, 7*(2), 131–174.

Hester, P. T., & Adams, K. M. (2014). *Systemic thinking—fundamentals for understanding problems and messes*. New York: Springer.

Houy, C., Fettke, P., & Loos, P. (2012). Understanding understandability of conceptual models—What are we actually talking about? In P. Atzeni, D. Cheung, & S. Ram (Eds.), *Conceptual modeling* (pp. 64–77). Berlin: Springer.

Huang, B., & Du, X. (2007). Analytical robustness assessment for robust design. *Structural and Multidisciplinary Optimization, 34*(2), 123–137.

Hwang, K. H., & Park, G. J. (2005). Development of a robust design process using a new robustness index. In *Proceedings of the ASME 2005 International Design Engineering Technical Conferences and Computers and Information in Engineering Conference* (Vol. 2: 31st Design Automation Conference, Parts A and B, pp. 231–241). New York: American Society of Mechanical Engineers.

IEEE, & ISO/IEC. (2010). IEEE and ISO/IEC Standard 24765: Systems and software engineering —Vocabulary. New York and Geneva: Institute of Electrical and Electronics Engineers and the International Organization for Standardization and the International Electrotechnical Commission.

ISO/IEC. (1991). ISO/IEC Standard 9126: Software product evaluation—Quality characteristics and guidelines for their use. Geneva: International Organization for Standardization and the International Electrotechnical Commission.

ISO/IEC. (2011). ISO/IEC Standard 25010: Systems and software engineering—Systems and software quality requirements and evaluation (SQuaRE)—System and software quality models. Geneva: International Organization for Standardization and the International Electrotechnical Commission.

Korczaka, E., & Levitin, G. (2007). Survivability of systems under multiple factor impact. *Reliability Engineering & System Safety, 92*(2), 269–274.

Kung, C. H. (1989). Conceptual modeling in the context of development. *IEEE Transactions on Software Engineering, 15*(10), 1176–1187.

Kung, C. H., & Solvberg, A. (1986). Activity modeling and behavior modeling of information systems. In T. W. Olle, H. G. Sol, & A. A. Verrijn-Stuart (Eds.), *Information system design methodologies: Improving the practice* (pp. 145–172). North Holland: Elsevier.

Lissitz, R. W., & Green, S. B. (1975). Effect of the number of scale points on reliability: A Monte Carlo approach. *Journal of Applied Psychology, 60*(1), 10–13.

Maxwell, K. (2001). The maturation of HCI: Moving beyond usability toward holistic interaction. In J. M. Carroll (Ed.), *Human-computer interaction in the new millennium* (pp. 191–209). New York: Addison-Wesley.

McCall, J. A., & Matsumoto, M. T. (1980). *Software quality measurement manual (RADC-TR-80-109-Vol-2)*. Griffiss Air Force Base, NY: Rome Air Development Center.

Mekdeci, B., Ross, A. M., Rhodes, D. H., & Hastings, D. E. (2011). Examining survivability of systems of systems. In *Proceedings of the 21st Annual International Symposium of the International Council on Systems Engineering* (Vol. 1, pp. 564–576). San Diego, CA: INCOSE-International Council on Systems Engineering.

Nakano, T., & Suda, T. (2007). Applying biological principles to designs of network services. *Applied Soft Computing, 7*(3), 870–878.

Nazir, M., & Khan, R. A. (2012). An empirical validation of understandability quantification model. *Procedia Technology, 4*, 772–777.

Nielsen, J. (1993). *Usability engineering*. Cambridge: Academic Press Professional.

Norman, D. A. (1999). Affordance, conventions, and design. *Interactions, 6*(3), 38–43.

Norman, D. A. (2013). *The design of everyday things* (Revised and expanded ed.). New York: Basic Books.

Ottensooser, A., Fekete, A., Reijers, H. A., Mendling, J., & Menictas, C. (2012). Making sense of business process descriptions: An experimental comparison of graphical and textual notations. *Journal of Systems and Software, 85*(3), 596–606.

Redman, J., Warren, M., & Hutchinson, W. (2005). System survivability: A critical security problem. *Information Management & Computer Security, 13*(3), 182–188.

Richards, M. G., Ross, A. M., Hastings, D. E., & Rhodes, D. H. (2008). Empirical validation of design principles for survivable system architecture. In *Proceedings of the 2nd Annual IEEE Systems Conference* (pp. 1–8).

Richards, M. G., Ross, A. M., Hastings, D. E., & Rhodes, D. H. (2009). Survivability design principles for enhanced concept generation and evaluation. In *Proceedings of the 19th Annual INCOSE International Symposium* (Vol. 2, pp. 1055–1070). San Diego, CA: INCOSE-International Council on Systems Engineering.

Ross, A. M., Rhodes, D. H., & Hastings, D. E. (2008). Defining changeability: Reconciling flexibility, adaptability, scalability, modifiability, and robustness for maintaining system lifecycle value. *Systems Engineering, 11*(3), 246–262.

Schach, S. R. (2002). *Object-oriented and classical software engineering* (5th ed.). New York: McGraw-Hill.

Wagner, S. (2013). *Quality models software product quality control* (pp. 29–89). Berlin: Springer.

Westrum, R. (2006). A typology of resilience situations. In E. Hollnagel, D. D. Woods, & N. Leveson (Eds.), *Resilience engineering: Concepts and precepts* (pp. 55–65). Burlington: Ashgate.

Zhang, P., Carey, J., Te'eni, D., & Tremaine, M. (2005). Integrating human-computer interaction development into the systems development life cycle: A methodology. *Communications of the Association for Information Systems, 15*, 512–543.

Chapter 12
Accuracy, Correctness, Efficiency, and Integrity

Abstract The design of systems and components during the design stage of the systems life cycle requires specific purposeful actions to ensure effective designs and viable systems. Designers are faced with a number of *other viability concerns* that they must embed into the design to ensure the system remains viable. The ability for a system to remain viable is critical if it is to continue to provide required functionality for its stakeholders. Other viability concerns includes the non-functional requirements for accuracy, correctness, efficiency, and integrity. Purposeful design requires an understanding of each of these requirements and how to measure and evaluate each as part of an integrated systems design.

12.1 Introduction to Accuracy, Correctness, Efficiency, and Integrity

This chapter will address four major topics (1) accuracy, (2) correctness, (3) efficiency, and (4) integrity. Each of these topics are associated with other viability concerns in design endeavors. The chapter begins by reviewing accuracy, its definitions, and concepts related to the reference value, precision, and trueness. The section will conclude with a discussion of how accuracy is approached as an aspect of purposeful systems design.

Section 12.2 will define correctness and demonstrate how both verification and validation activities provide evaluation opportunities to ensure correctness. The section will also include four design principles that support the development of systems that correctly represent the specified requirements for the system being addressed by the design team.

Section 12.3 in this lecture will address efficiency by providing a clear definition for efficiency and establishes a proxy for system efficiency. The section will conclude by detailing a generic attribute and method for evaluating efficiency in systems design endeavors.

K.M. Adams, *Non-functional Requirements in Systems Analysis and Design*,
Topics in Safety, Risk, Reliability and Quality 28,
DOI 10.1007/978-3-319-18344-2_12

Section 12.4 will define integrity and the concept that underlies its use as a non-functional requirements in systems designs. The section will also discuss 33 security design principles, and the life cycle stages where they should be invoked when designing for systems for integrity.

Section 12.7 will define a measure and a means for measuring other viability concerns that is a function of accuracy, correctness, efficiency, and integrity. The section will conclude by relating the proposed measure for other viability concerns as a metric and will include a structural map for accuracy, correctness, efficiency, and integrity.

The chapter has a specific learning goal and associated objectives. The learning goal of this chapter is to be able to understand the other viability concerns and to identify how the non-functional requirements of accuracy, correctness, efficiency, and integrity affect design in systems endeavors. This chapter's goal is supported by the following objectives:

- Define accuracy.
- Discuss how accuracy is achieved during purposeful systems design.
- Define correctness.
- Describe how verification and validation are related to correctness.
- Describe the design principles that may be used to support correctness in design endeavors.
- Define efficiency.
- Discuss how resources may be used as a proxy for efficiency in systems design endeavors.
- Define integrity.
- Discuss some of the 33 security design principles associated with integrity is systems design endeavors.
- Construct a structural map that relate other viability concerns to a specific metric and measureable characteristic.
- Explain the significance of accuracy, correctness, efficiency, and integrity in systems design endeavors.

The ability to achieve these objectives may be fulfilled by reviewing the materials in the sections that follow.

12.2 Accuracy

In this section the basics of accuracy and how it is applied during systems endeavors will be discussed. The definition for the non-functional requirement termed accuracy would, on the surface, seem to be straight forward. However, this very common term is routinely used incorrectly and is often misrepresented as precision. To improve understanding of accuracy and to use this term as it was intended, both a common definition and associated terminology will be constructed and a detailed graphic will be used to represent a concept for accuracy. This is an important first step if accuracy is to be understood as an element of systems design endeavors.

12.2.1 Definition for Accuracy

Accuracy, from a systems engineering perspective, is defined as:

> 1. A qualitative assessment of correctness, or freedom from error. 2. A quantitative measure of the magnitude of error. (IEEE and ISO/IEC 2010, p. 6)

Accuracy has additional definitions, shown in Table 12.1. that may provide further meaning for the term when applied as a non-functional requirement for a system.

Table 12.1 Additional definitions for accuracy

Definition	Source
"The measurement of the degree to which a given measurement may deviate from the truth"	Churchman and Ratoosh (1959, p. 92)
"Accuracy is a characteristic of the measurement result itself"	De Bièvre (2006, p. 645)
"A qualitative performance characteristic expressing the closeness of agreement between a measurement result and the value of the measurand"	Menditto et al. (2007, p. 46)
"The general meaning of accuracy is the proximity of a value or a statistic to a reference value. More specifically, it measures the proximity of the estimator T of the unknown parameter θ to the true value of θ. The accuracy of an estimator can be measured by the expected value of the squared deviation between T and θ, in other words: $E[(T-\theta)^2]$" Accuracy should not be confused with the term precision, which indicates the degree of exactness of a measure and is usually indicated by the number of decimals after the comma	Dodge (2008, p. 1–1)
"Closeness of agreement between a measured quantity value and a true quantity value of a measurand NOTE 1 The concept 'measurement accuracy' is not a quantity and is not given a numerical quantity value. A measurement is said to be more accurate when it offers a smaller measurement error NOTE 2 The term "measurement accuracy" should not be used for measurement trueness and the term "measurement precision" should not be used for 'measurement accuracy', which, however, is related to both these concepts NOTE 3 'Measurement accuracy' is sometimes understood as closeness of agreement between measured quantity values that are being attributed to the measurand"	JCGM (2012, p. 21)
"Reflects the closeness between the measurement result and the true value of the measurand"	Rabinovich (2013, p. 2)

These definitions may be adequate for metrologists and scientists who engage in measurement activities on a daily basis. However, for those engaged in engineering and design activities, the definitions in Table 12.1 require additional description if accuracy is to be properly invoked as a meaningful non-functional requirement in a design endeavor. The next section will provide additional context for the term accuracy in support of a common meaning.

12.2.2 Accuracy in Measurement

The discussion of accuracy must be placed in context. The context in which accuracy will be addressed is that of testing and the associated elements which include the measurement method and the larger measurement process.

12.2.2.1 Measurement Method

A measurement method "consists of a prescription or written procedure by which one can go about the business of taking measurements on the properties of some physical material" (Murphy 1969, p. 357). The measurement or test method is a formal element specified in the systems design.

12.2.2.2 Measurement Process

The measurement process includes a number of elements that include the: "(a) measurement method, (b) a system of causes, (c) repetition, and (d) capability of control" (Murphy 1969, p. 357). The measurement process is how the measurement method is implemented. The system of causes includes the resources required to execute the test (i.e., materials, test personnel, instruments, test environment, and specific time). An important element is the notion of control. The test must be capable of statistical control if it is included as part of a formal measurement process. Without the presence of statistical control the process cannot use the routine statistical measures that permit us to discuss both accuracy and precision.

12.2.2.3 Accuracy as a Qualitative Performance Characteristic

All measurement and testing occurs with respect to a *reference level* or *target value*. The target value is most often established as either (1) a property of a material or (2) a physical characteristic of a system component. The target value serves as the design value against which we will measure during the conduct of a test. Accuracy connotes the agreement between the long-run average of the actual measurements and the target value. Accuracy is "a qualitative performance characteristic expressing the closeness of agreement between a measurement result and the value of the measurand" (Menditto et al. 2007). The qualitative performance characteristic of the measurement, which is accuracy, includes both *precision* and *trueness*.

12.2.2.4 Accuracy: Precision and Trueness

Accuracy is described by both precision and trueness. The definitions for each term are as follows:

- *Precision*: "Closeness of agreement between indications or measured quantity values obtained by replicate measurements on the same or similar objects under specified conditions" (JCGM 2012, p. 22)
- *Trueness*: "Closeness of agreement between the average of an infinite number of replicate measured quantity values and a reference quantity value" (JCGM 2012, p. 21).

Figure 12.1 is a depiction of how precision and trueness are used to describe the accuracy of a measurement context. From both the depiction in Fig. 12.1 and a knowledge of basic statistics, it is clear that the "standard deviation is an index of precision" (Murphy 1969, p. 360). The process improvement notion of *Six Sigma*, popularized by efforts at Motorola (Hahn et al. 1999) is based upon a precision that is a multiple of six standard deviations (6σ) which signifies a process which is said to be *under control*, having a minuscule rejection rate of 0.00034 % with an associated yield of 99.99966 %.

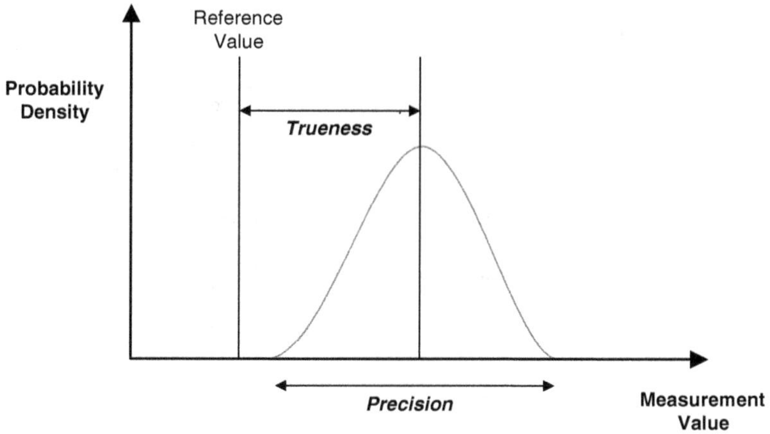

Fig. 12.1 Precision and trueness in a measurement context

12.2.2.5 Accuracy, Precision, Trueness and Error

There is a relationship between the types of errors present during measurement and testing endeavors and the associated performance characteristics represented by accuracy, precision, and trueness.

- *Trueness Errors*: The difference between the mean of the measurement process and the reference value is termed a systematic error and is expressed by a quantitative value we term *bias*.
- *Precision Errors*: The measurements that contribute to the mean and exist within some index of precision are caused by random errors and are expressed by a quantitative value we term the standard deviation.
- *Accuracy Errors*: The total error encountered during the measurement process, attributable to both systematic and random errors and is expressed as measurement uncertainty.

The relationship between these types of errors, performance characteristics, and the quantitative expression is depicted in Fig. 12.2. It is important to recognize that the accuracy in a measurement is a parameter that expresses the measurement uncertainty or more precisely, "the dispersion of the values that could reasonably be attributed to the measurand" (i.e. the quantity to be measured) (Menditto et al. 2007, p. 46).

> Measurement uncertainty conveys more correctly the slight doubt which is attached to any measurement result. Thus a doubtful meaning of 'accuracy' (doubtful because tied to 'true value') is replaced by a practical one: 'measurement uncertainty'. (De Bièvre 2006, p. 645)

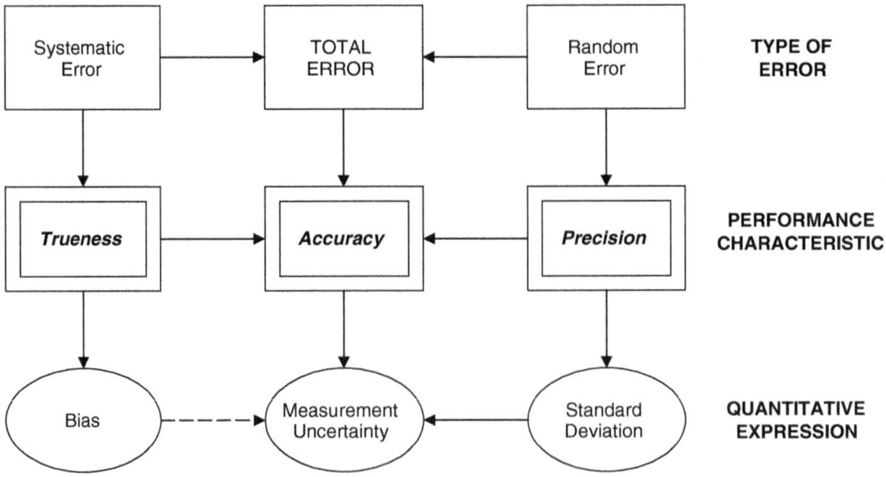

Fig. 12.2 Relationships between type of error, performance characteristic and quantitative expression

Finally, it is important to note that the precision of a measurement process is related to both repeatability and reproducibility. Repeatability addresses short-time repetition of the measurement method by the same test personnel while reproducibility addresses the measurement process by a different group of test personnel that applies the same measurement method.

12.2.3 Accuracy in Systems Design

In this section the specific tasks and activities accomplished during the design stage of the systems life cycle that require accuracy as an element of the performance specification will be identified. Accuracy is an important non-functional requirement where measures of effectiveness (MOE), measures of performance (MOP), and technical performance measures (TPM) are required. Accuracy is a factor in determining system performance requirements and is addressed in *IEEE Standard 1220—Systems engineering—Application and management of the systems engineering process* (IEEE 2005) specifically in the following areas.

12.2.3.1 Conceptual Design Stage Activities Where Accuracy is a Consideration

Section 5.1.1.3 in IEEE Standard 1220 (IEEE 2005) requires the following:

> Identify the subsystems of each product and define the design and functional interface requirements among the subsystems and their corresponding performance requirements and design constraints. System product functional and performance requirements should be allocated among the subsystems so as to assure requirement traceability from the system products to their respective subsystems, and from subsystems to their parent product. (IEEE 2005, pp. 21–22)

The design team develops top-level performance measures that are labeled Measures of Effectiveness (MOE). MOEs are defined as:

> Standards against which the capability of a solution to meet the needs of a problem may be judged. The standards are specific properties that any potential solution must exhibit to some extent. MOEs are independent of any solution and do not specify performance of criteria. (Sproles 2001, p. 146)

MOEs have two key characteristics: (1) the ability to be tested, and (2) that they can be quantified in some manner. For example, if a team is tasked with designing an electric vehicle a valid MOE may be stated as: The electric vehicle must be able to drive fully loaded from Norfolk, VA to Washington, DC without recharging. This MOE is clearly measurable and quantifiable. MOEs are typically supported by a supporting lower-level hierarchy of Measures of Performance (MOP).

12.2.3.2 Preliminary Design Stage Activities Where Accuracy is a Consideration

Section 5.2.1.1 in IEEE Standard 1220 (IEEE 2005) requires the following:

> Subsystem performance requirements are allocated among the assemblies so as to assure requirements traceability from subsystems to appropriate assemblies, and from assemblies to the parent subsystem. (IEEE 2005, p. 25)

Section 5.2.1.2 in IEEE Standard 1220 (IEEE 2005) requires the following:

> Assembly performance requirements are allocated among the components so as to assure requirement traceability from the assemblies to their respective components, and from components to their parent assembly. (IEEE 2005, p. 25)

The design team develops mid-level performance measures that are labeled Measures of Performance (MOP). MOPs are defined as:

> Performance requirements describe how well functional requirements must be performed to satisfy the MOEs. These performance requirements are the MOPs that are allocated to subfunctions during functional decomposition analysis and that are the criteria against which design solutions [derived from synthesis (see 6.5)] are measured. There are typically several MOPs for each MOE, which bind the acceptable performance envelope. (IEEE 2005, p. 41)

In this case the performance requirement for the electric vehicle that was used as an example during the establishment of the type of MOE in the earlier conceptual design phase can be clearly traced. In the example the MOE stated that the electric vehicle must be able to drive fully loaded from Norfolk, VA to Washington, DC without recharging. In support of this requirement, more than one MOP may be developed. An example of a supporting MOP is: the vehicle range must be equal to or greater than 250 miles. This establishes a more precise requirement than the distance from Norfolk, VA to Washington, DC. MOPs are supported by any number of specific Technical Performance Measures (TPMs).

12.2.3.3 Detailed Design Stage Activities Where Accuracy is a Consideration

Section 5.3.4.1 in IEEE Standard 1220 (IEEE 2005) requires the following:

> Component reviews should be completed for each component at the completion of the detailed design stage. The purpose of this review is to ensure that each detailed component definition is sufficiently mature to meet measure of effectiveness/measure of performance (MOE/MOP) criteria. (IEEE 2005, p. 31)

Section 6.1.13 in IEEE Standard 1220 (IEEE 2005) requires the following:

> Identify the technical performance measures (TPMs), which are key indicators of system performance. Selection of TPMs are usually limited to critical MOPs that, if not met, put the project at cost, schedule, or performance risk. Specific TPM activities are integrated into the SEMS [Systems Engineering Master Schedule] to periodically determine achievement to date and to measure progress against a planned value profile. (IEEE 2005, p. 42)

The design team defines detailed Technical Performance Measures (TPMs) for all of the MOPs associated with the systems requirements. TPMs are quantitative in nature and are derived directly from and support the mid-level MOPs. The TPMs are used to assess compliance with requirements in the system's requirements breakdown structure (RBS) and also assist in monitoring and tracking technical risk.

In the previous example for an electric vehicle the MOE stated that the electric vehicle must be able to drive fully loaded from Norfolk, VA to Washington, DC without recharging. In support of this MOE an MOP was developed and stated the vehicle range must be equal to or greater than 250 miles. This established a more precise requirement than the distance from Norfolk, VA to Washington, DC. In support of this a series of specific Technical Performance Measures (TPMs) are invoked. The TPMs for this example may include performance measures such as: battery capacity, vehicle weight, drag, power train friction, etc. Each of the TPMs must have a measurement accuracy—which requires specification for a reference value, precision, and trueness for each and every measure.

In conclusion, accuracy is a purposeful design function that permits systems to be tested and evaluated against standard reference values throughout the entire systems' life.

The section that follows will address the non-functional requirement for correctness.

12.3 Correctness

In this section, the basics of correctness and how it is applied during systems endeavors will be reviewed. As with many of the other non-functional requirements, correctness sounds like a very clear concept. However, arriving at a common definition and understanding for how correctness is invoked during requirements deliberations in systems design endeavors may turn out to be problematic. Once again, take a minute and review the index of a favorite systems engineering or software engineering text and look for the word *correctness*. Is it missing? Once again, it would not be surprising to hear that the word is missing from just about every major text. Therefore, a careful review of correctness and its characteristics is in order to provide a common base for both learning and application during systems design endeavors.

12.3.1 Definition for Correctness

Correctness, from a systems engineering perspective, is defined as:

> The degree to which a system or component is free from faults in its specification, design, and implementation. (IEEE and ISO/IEC 2010, p. 81)

Table 12.2 Additional definitions for correctness

Definition	Source
"Extent to which a program satisfies its specifications and fulfills the user's mission objectives"	Cavano and McCall (1978, p. 136)
"Extent to which a program satisfies its specifications and fulfills the user's mission objectives"	McCall and Matsumoto (1980, p. 12)
"How well does it conform to the design requirements?"	Bowen et al. (1985, pp. 2–1)
"1. Program satisfies specification; 2. Meets user expectations; 3. Fault-free"	Blundell et al. (1997, p. 236)

Correctness has additional definitions, from the literature, that are listed in Table 12.2 that may provide further help in understanding the term when applied as a non-functional requirement for a system.

From these definitions, correctness is *the degree to which a system satisfies its specified design requirements*. Having settled on this basic definition, the next section will discuss how correctness is evaluated during systems design endeavors.

12.3.2 Evaluating Correctness in Systems Designs

The definition chosen for correctness, *the degree to which a system satisfies its specified design requirements,* seems to sound very much like the process requirements for verification and validation activities that are performed throughout the systems life cycle. As a result, a review of the definitions for both verification and validation will be conducted.

12.3.2.1 Verification

Three definitions for verification are presented in Table 12.3.

From the definitions in Table 12.3 verification is concerned with ensuring that requirements have been satisfactorily met. In all of the design stages and in the

Table 12.3 Definitions for verification

Definition	Source
"The process of evaluating a system or component to determine whether the products of a given development phase satisfy the conditions imposed at the start of that phase"	IEEE (2012, p. 9)
"Confirmation, through the provision of objective evidence, that specified requirements have been fulfilled"	IEEE and ISO/IEC (2008, p. 8)
"Confirmation by examination and provision of objective evidence that the specified requirements to which an end product is built, coded, or assembled have been fulfilled"	ANSI/EIA (1998, p. 64)

construction stage, verification examines the result of a given process or activity to determine conformity with the stated requirement. During the design stages this is conformance to the functional, technical, or interface requirements. During the construction stage this may include conformance with specific physical or operational parameters such as length, voltage, resistance, strength, etc. Now that verification is defined, a review of validation will be conducted.

12.3.2.2 Validation

Three definitions for validation are presented in Table 12.4.

From the definitions in Table 12.4 validation is concerned with ensuring that the intended end-use for the system has been met.

For many, the difference between the terms verification and validation, and in this case the activities performed, is hard to differentiate based on the formal definitions contained in Tables 12.3 and 12.4. So, it may be easier to think of these terms using the following:

- *Validation*—Is the system doing the right job? Here the concern is with ensuring that the system that was designed or constructed is doing what it was intended to do.
- *Verification*—Is the system doing the job right? Here the concern is with ensuring that the design or constructed item meets the specified requirements.

The activity that assures the system's requirements capture the stakeholder's intentions is referred to as validation while the activity of assuring that the system meets its specified requirements is referred to as verification. In other words, validation tasks are the activities that link the system's requirements to the intention, while verification tasks link the system's requirements to the actual design and its implementation. In the next section a review of how designers purposefully address correctness as part of the design process will be presented.

Table 12.4 Definitions for validation

Definition	Source
"The process of evaluating a system or component during or at the end of the development process to determine whether it satisfies specified requirements"	IEEE (2012, p. 9)
"Confirmation, through the provision of objective evidence, that the requirements for a specific intended use or application have been fulfilled"	IEEE and ISO/IEC (2008, p. 8)
"Confirmation by examination and provision of objective evidence that the specific intended use of an end product (developed or purchased), or an aggregation of end products, is accomplished in an intended usage environment"	ANSI/EIA (1998, p. 64)

12.3.3 Methods for Ensuring Correctness During System Design

Now that correctness, verification, and validation are defined and an understanding of the activities used to analyze correctness as part of the design effort have been established, how designers ensure correctness during design activities will be addressed.

Systems design activities must satisfy the specified requirements for the system. Correctness of design means that the design is sufficient. A sufficient design is one that satisfactorily incorporates the elements of the *System Theory's Design Axiom* (Adams et al. 2014). This is accomplished by integrating the *Design Axiom's* supporting principles into the design rules used by the design team. The four supporting principles from the Design Axiom are addressed in the following sections.

12.3.3.1 Law of Requisite Parsimony

The Law of Requisite Parsimony states that human beings can only deal simultaneously with between five and nine observations at one time. This is based on George Miller's seminal paper *The Magical Number Seven, Plus or Minus Two: Some Limits on Our Capacity for Processing Information* (1956). Miller makes three points, based on empirical experimentation, in this paper.

1. *Span of attention*: Experiments showed that when more than seven objects were presented, the subjects were said to estimate and for less than seven objects they were said to subitize. The break point was at the number seven.
2. *Span of immediate memory*: He reports the fact that the span of immediate memory, for a variety of test materials, is about seven items in length.
3. *Span of absolute judgment*: This is the clear and definite limit based on the accuracy with which we can identify absolutely the magnitude. This is also in the neighborhood of seven.

Nobel Laureate Herbert A. Simon [1916–2001] followed Miller's work with a paper (Simon 1974) that questioned Miller's vagueness with respect to the size of a chunk used in memory and is a worthy companion to Miller's work. Application of the Law of Principle of Requisite Parsimony during systems design endeavors is clear.

> No matter how complex our models, designs, or plans may be, we should avoid subjecting people to more than nine concepts simultaneously, and should as a matter of routine, involve fewer. Parsimony should be invoked in systems engineering in order to make sure that the artefacts and methods we use in designing systems does not inherently try to force people to make judgments that exceed their short term cognitive capacities. (Adams 2011, p. 149)

The ability to conceptualize, design-to, or implement ideas that have more than 9 individual elements strains normal human memory and recall. Design artifacts, systems hierarchies, and organizations should adopt structures that utilize spans of less than 9. This application of parsimony improves the ability to conceptualize, design, and implement systems that are correct.

12.3.3.2 Law of Requisite Saliency

Economist Kenneth Boulding [1910–1993], identified three reasons for poor intellectual productivity (1966):

1. *Spurious saliency*: emphasizing the wrong things out of proportion to what they deserve.
2. *Unproductive emulation*: behaving like those who help create rather than resolve problems.
3. *Cultural lag*: not using established knowledge with dispatch.

The significance of Boulding's impact on design is that not all design elements have the same level of importance.

> The design process must incorporate specific provision for uncovering the relative saliency of the factors in the Design Situation and the factors that characterize the Target, in order to achieve the kind of understanding that is needed to put the design factors in proper perspective. (Warfield 1999, p. 34)

The design team must make numerous decisions and in doing so some elements of the design will take priority over others. Requisite saliency has particular importance to the design team because as they conduct trade-off analyses, solve problems, and process data and information into knowledge, requisite saliency provides the means for making these rational decisions. As a result, all design "processes must include a specific provision for uncovering relative saliency for all the factors in a system as a critical element in the overall system design" (Adams 2011, p. 149).

12.3.3.3 Minimum Critical Specification

The father of modern socio-technical systems design, Albert Cherns, has spent a major portion of his career emphasizing the need to limit the scope of the design process to that which is required, no more and no less. Failure to limit the process in this manner leads to terms such as *brass-plating* and *polishing-the-cannonball*, which serve to indicate that objects or systems have been far over designed and include elements never envisioned by the system's stakeholders. Cherns' principle of minimum critical specification states:

> This principle has two aspects, negative and positive. The negative simply states that no more should be specified than is absolutely essential; the positive requires that we identify what is essential. (Cherns 1987, p. 155)

The basic utility of this principle is clear—design as little as possible and only specify what is essential. However, all design require redundancy to effect many of the non-functional requirements described in this book. The design becomes a balance between all of the system's requirements and the resources (i.e., primarily financial) allocated to implement the design. Cherns expressed concern over the tendency for design teams to focus on a potential solution too early in the design process, thereby closing options before the team could rationally evaluate all alternatives.

> This premature closing of options is a pervasive fault in design; it arises, not only because of the desire to reduce uncertainty, but also because it helps the designer to get his own way. We measure our success and effectiveness less by the quality of the ultimate design than by the quantity of our ideas and preferences that have been incorporated into it. (Cherns 1987, p. 155)

The ability to balance the specifications for the design against the known requirements is most often a function of incomplete knowledge of the system (i.e., the principle of darkness), which only improves as we improve our understanding of the system during analysis.

> Whatever benefits we plan to achieve through specification become obsolete (often at a rapid pace) as the contextual elements surrounding the design become better defined. In many cases, early over specification may have a crippling effect on the ability of the design team to adapt to evolving changes in context. (Adams 2011, p. 150)

12.3.3.4 Pareto Principle

The Pareto Principle was named after the 19th century Italian economist Vilfredo Pareto [1848–1923], who noticed that in Italy about 80 % of wealth was in the hands of 20 % of the population. Since that time a variety of sociological, economic, and political phenomena have been shown to have the same pattern. The well-known statistical quality control expert Joseph M. Juran [1904–2008] "… claims credit for giving the Pareto Principle its name. Juran's Pareto Principle is sometimes known as the Rule of 80/20" (Sanders 1987, p. 37).

> The Pareto Principle states that in any large complex system 80 % of the output will be produced by only 20 % of the system. The corollary to this is that 20 % of the results absorb 80 % of the resources or productive efforts. (Adams 2011, p. 147)

This fairly simple principle can be utilized during the design process by understanding that some system-level requirements will consume more design resources (e.g., time, manpower, methods, etc.) than others. Similarly, elements or components of the system may consume more power, require more information, process more data, or fail more often than others. By using the Pareto principle, and its accompanying ratios, the design team may be able to better understand the relationships and between the system and its constituent elements.

Table 12.5 Design principles invoked to support correctness

Design principle	Description	Source
Law of requisite parsimony	Human short-term memory is incapable of recalling more than seven plus or minus two items	Miller (1956) and Simon (1974)
Law of requisite saliency	The factors that will be considered in a system design are seldom of equal importance. Instead, there is an underlying logic awaiting discovery in each system design that will reveal the saliency of these factors	Boulding (1966) and Warfield (1999)
Principle of Minimum critical specification	This principle has two aspects, negative and positive. The negative simply states that no more should be specified than is absolutely essential; the positive requires that we identify what is essential	Cherns (1976, 1987)
Pareto principle	80 % of the objectives or outcomes are achieved with 20 % of the means	Bresciani-Turroni (1937), Creedy (1977) and Sanders (1987)

12.3.4 Summary for Correctness

In summary, systems designers have a number of methods and techniques at their disposal that support the engineering of systems that correctly represent the specified requirements. Table 12.5 provides a list of the four supporting principles from the *System's Theory Design Axiom* (Adams et al. 2014) that can be implemented by the design team, as specific design rules, that can help ensure that designs are correct. At the completion of specific tasks and activities in the design stages both verification and validation activities ensure that the system's design is satisfactory through comparison with the system's specified requirements. All of the factors discussed in this section must be addressed by the systems designer to ensure correctness of requirements as part of the purposeful design during systems endeavors. The section that follows will address system efficiency.

12.4 Efficiency

In this section, the non-functional requirement for efficiency will be reviewed as an additional viability concern to be considered during systems design endeavors. As the last non-functional requirement to be considered, efficiency is a requirement closely related to system performance. Efficiency, like productivity, is a primary measure used in the determination of performance and includes both economic and production perspectives. Because efficiency is such an important viability concern, this requirement will be approached in the sections that follow by reviewing its formal definition, discussing concepts that surround its utilization, and how it is treated during systems design endeavors.

12.4.1 Definition for Efficiency

Efficiency, from a systems engineering perspective, is defined as:

> 1. The degree to which a system or component performs its designated functions with minimum consumption of resources. 2. Producing a result with a minimum of extraneous or redundant effort. (IEEE and ISO/IEC 2010, p. 120)

Efficiency has additional definitions, from the literature, that are listed in Table 12.6 that may provide further help in understanding the term when applied as a non-functional requirement for a system.

Table 12.6 Additional definitions for efficiency

Definition	Source
"The extent that it fulfills its purpose without waste of resources"	Boehm et al. (1976, p. 604)
"The amount of computing resources and code required by a program to perform a function"	Cavano and McCall (1978, p. 136)
"The software does not meet performance (speed, storage) requirements. The rating is in terms of effort required to modify software to meet performance requirements"	McCall and Matsumoto (1980, p. 28)
"Efficiency is a concept based on the physical and engineering sciences and concerns the relationship between *inputs* and *outputs*"	Szilagyi (1984, p. 30)
"Efficiency is a measure of the use of effective capacity in producing a particular result such as a part or a product"	Del Mar (1985, p. 128)
"Efficiency deals with utilization of a resource. The rating formula for efficiency is in terms of actual utilization of a resource and budgeted allocation for utilization. For example, if a unit is budgeted for 10 % available memory and actually uses 7 %, the rating formula shows an efficiency level of 0.3 $(1 - 0.07/0.10 = 0.3)$"	Bowen et al. (1985, pp. 3–5)
"On the first level is physical efficiency expressed as output divided by inputs of such physical units as Btu's, kilowatts, and foot-pounds. On the second level are economic efficiencies. These are expressed in term s of economic units of output divided by economic units of input, each expressed in terms of medium of exchange such as money"	Thuesen and Fabrycky (1989, p. 6)
"Efficiency, that is, the ability to do things right, rather than the ability to get the right things done"	Drucker (2001, p. 192)

From these definitions, efficiency is *a measure of a systems utilization of resources.* It is important to note that resources include six categories represented by the acronym M5I: (1) material, (2) manpower, (3) money, (4) minutes or time, (5) methods, and (6) information. Armed with this comprehensive definition, a review of efficiency and the concepts that surround both its utilization and how it is may be used as a purposeful element during systems design endeavors will be conducted.

12.4.2 Addressing System Efficiency During Design Endeavors

In the section that follows a proxy for system efficiency and a detailed generic attribute for evaluating efficiency in systems design endeavors are presented.

12.4.2.1 System Efficiency in Engineering

Efficiency, from an engineering perspective, is generally viewed as a performance characteristic where the outputs are evaluated against the inputs. "The engineer must be concerned with two levels of efficiency … physical efficiency and economic efficiency" (Thuesen and Fabrycky 1989, p. 6). These concerns are related by the simple relationships in Eqs. 12.1 and 12.2.

Physical Efficiency (Thuesen and Fabrycky 1989, p. 6)

$$\mathit{Efficiency}\,(\mathit{physical}) = \frac{\mathit{output}}{\mathit{input}} \tag{12.1}$$

Economic Efficiency (Thuesen and Fabrycky 1989, p. 6)

$$\mathit{Efficiency}\,(\mathit{economic}) = \frac{\mathit{worth}}{\mathit{cost}} \tag{12.2}$$

When efficiency is viewed from the perspective of an engineering design team, it is clear that they are faced with the need to clearly address efficiency using design principles related to the problem at hand. The design team develops heuristics, analogies, and metaphors that help them during the development and evaluation of a series of design options.

> The design problem characteristics are used to actively seek and exploit appropriate analogies and metaphors to generate new options and facilitate and simplify systems architecting and implementation. Thereafter, the candidate design options are evaluated using proxies for elegance metrics (i.e., effectiveness, robustness, efficiency, predictability, and implementation ease). The proxies for effectiveness are usability/operability and perceived utility. The proxies for robustness are physical and functional redundancy. The proxies for efficiency are number of steps to complete task and resource utilization. (Madni 2012, p. 352)

The proxy for efficiency—*the number of steps to complete task and resource utilization*, is a valid method to evaluate efficiency in a systems design. Systems expert Russell Ackoff [1919–2009] also related efficiency to resources (M5I) stating:

> Information, knowledge, and understanding enable us to increase efficiency, not effectiveness. The efficiency of behavior or an act is measured relative to an objective by determining either the amount of resources required to obtain that objective with a specified probability, or the probability or obtaining that objective with a specified amount of resources. (Ackoff 1999, p. 171)

In a formally defined engineering design process (which is a system), the process that is clearly defined and orderly requires less energy to execute than one that is poorly defined and disorganized. A poorly defined and disorganized process inefficiently utilizes resources and ends up "diverting energy to exploring new paths (thereby wasting energy and reducing efficiency)" (Colbert 2004, p. 353). Based upon this analysis, resource efficiency may serve as a valid proxy for systems design efficiency.

12.4.2.2 Evaluating Efficiency in System Design Endeavors

In 1999 the Electronic Industries Association (EIA) issued interim standard 731-1, the Systems Engineering Capability Model or SECM (EIA 1999).

> The SECM was a merger of two previous widely used systems engineering models: the Systems Engineering Capability Maturity Model® (SE-CMM®) and the Systems Engineering Capability Assessment Model (SECAM). The EIA completed this effort and published the SECM version 1.0 as an Interim Standard in January 1999. This document was then used as the main systems engineering source document for the CMMI℠ development. (Minnich 2002, p. 62)

The SECM contains general attributes (GA) for both usefulness and cost effectiveness. The cost effectiveness GA is defined as "the extent to which the benefits received are worth the resources invested. Cost effectiveness is determined through the use of an intermediate parameter—resource efficiency" (Wells et al. 2003, p. 304). The GA for cost effectiveness turns out to be a useful measure for the efficiency of the design process. The resource efficiency calculation is a ratio of the actual resources required to produce results against the benchmarked standard, as depicted in Eq. 12.3.

Resource Efficiency

$$Resource\ Efficiency = \frac{actual\ resources}{benchmarked\ standard} \tag{12.3}$$

A five-point Likert scale may be utilized to evaluate efficiency as depicted in Table 12.7 (Wells et al. 2003).

The next section will discuss how integrity is addressed in systems design endeavors.

Table 12.7 Resource Efficiency Likert Scale [Adapted from Wells et al. (2003, Table II)]

Measure	Descriptor	Measurement criteria
0.0	E−− (E minus, minus)	Resources required to produce the work product(s) or result(s) exceeded the expected (benchmarked) values by more than 50 %
0.5	E− (E minus)	Resources required to produce the work product(s) or result(s) were more than the expected (benchmarked) values by 5–50 %
1.0	E	Resources required to produce the work product(s) or result(s) were within 5 % of the expected (benchmarked) values
1.5	E+ (E plus)	Resources required to produce the work product(s) or result(s) were less than the expected (benchmarked) values by 5–50 %
2.0	E++ (E plus, plus)	Resources required to produce the work product(s) or result(s) were less than the expected (benchmarked) values by more than 50 %

12.5 Integrity

In this section the basics of system integrity and how it is applied during systems endeavors is reviewed Integrity, as a non-functional requirement, is a term that is very rarely used during discussions about systems requirements. Because of its infrequent use in ordinary conversation, the formal systems vocabulary definition as well as some definitions from the literature will be reviewed. This will solidify a common meaning for the term during discussions of its use during systems design endeavors.

12.5.1 Definition for Integrity

Integrity, from a systems engineering perspective, is defined as:

> The degree to which a system or component prevents unauthorized access to, or modification of, computer programs or data. (IEEE and ISO/IEC 2010, p. 181)

Integrity's additional definitions are taken from the literature and are listed in Table 12.8. All of these definitions may be used to better understand the term when applied as a non-functional requirement for a system.

When the definitions in Table 12.8 are reviewed it is clear that, over time, the definition for integrity has taken on new meaning. From these shifts and interpretations the integrity of a system can be taken to mean *the systems' ability to ensure program correctness, noninterference, and information assurance*. Integrity is concerned with information modification rather than information disclosure or availability. That is, integrity is something different from confidentiality or denial of service.

Table 12.8 Additional definitions for integrity

Definition	Source
"The maintenance of information validity"	Biba (1975, p. 55)
"Ensuring that data are altered only in an approved fashion"	Jacob (1991, p. 90)
"Is concerned with improper modification of information or processes"	Sandhu and Jajodia (1993, p. 482)
"An object has integrity if its quality meets expectations which were determined before the fact"	Courtney and Ware (1994, p. 208)
"Being concerned with the improper modification of information (much as confidentiality is concerned with improper disclosure). We understand modification to include insertion of new information and deletion of existing information, as well as changes to existing information"	Sandhu and Jajodia (1994, p. 617)
"A system's ability to avoid undesirable alteration due to the presence of errors"	Bowen and Hinchey (1998, p. 661)
"One attribute of *dependability*, that is, integrity is defined as *dependability with respect to absence of improper alterations.* Another interpretation of integrity is that the system concerned provides *external consistency*, that is, the *correct correspondence between data objects and the real world*"	Foley (2003, pp. 37–38)
"(i) Protecting against unauthorized modification of information, (ii) protecting against unintentional modification of information caused by system failures or user errors"	Georg et al. (2003, p. 42)
"Integrity has an important difference from confidentiality: a computing system can damage integrity without any external interaction, simply by computing data incorrectly. Thus, strong enforcement of integrity requires proving program correctness"	Sabelfeld and Myers (2003, p. 7)
"Program correctness, noninterference, and data invariant conditions, which yields the following categorization: – Program correctness: program output is precise, accurate, meaningful and correct with respect to a specification – Noninterference: data is modified only by authorized people or processes either directly, or indirectly by means of information flow – Data invariants: data is precise or accurate, consistent, unmodified, or modified only in acceptable ways under program execution"	Li et al. (2003, p. 48)
"The security goal that generates the requirement for protection against either intentional or accidental attempts to violate data integrity (the property that data has not been altered in an unauthorized manner) or system integrity (the quality that a system has when it performs its intended function in an unimpaired manner, free from unauthorized manipulation)"	Stoneburner et al. (2004, p. A-2)

12.5.2 Integrity Concepts

In order to remain viable, systems must maintain integrity at all times. The system's stakeholders must have a high degree of trust that their system and its attendant data and information are correct (i.e., precise, accurate, and meaningful), valid (created, modified and deleted only by authorized users), and invariant (i.e., consistent and unmodified).

The National Institute for Standards and Technology (NIST) has issued its *Engineering Principles for Information Technology Security* (Stoneburner et al. 2004) and in this publication they define system security as "The quality that a system has when it performs its intended function in an unimpaired manner, free from unauthorized manipulation of the system, whether intentional or accidental" (p. A-4). During the engineering stages of the systems life cycle the design team is tasked with ensuring that the system's design includes both requirements and means to ensure adequate system integrity. There are 33 security principles that may be used to enhance system integrity during system design endeavors. The section that follows will discuss how each of these security principles may be applied to deliver integrity as a purposeful element of the system design process.

12.5.3 Integrity in Systems Design

In order to successfully, incorporate considerations that will sufficiently address system integrity, six categories of security-related design principles will be addressed as an element of purposeful system design. Table 12.9 displays the six categories, the 33 security design principles, and the five major systems life cycle stages where the principles may be applied. The major system life cycle stages are: (1) conceptual design [C], (2) design and construction [D&C], (3) test and evaluation [T&E], (4) operations and maintenance [O&M], and (5) retirement and disposal [R&D].

A discussion of each of the 33 security principles, is contained in the *NIST Special Publication 800-27—Engineering Principles for Information Technology Security* (Stoneburner et al. 2004).

By focusing on the appropriate security principle during the temporal design stage, the system's designers ensure that integrity considerations are invoked as a purposeful element of the system's design. By basing understanding on the 33 established security principles the designers are able to create designs that meet the requirements for systems integrity during the appropriate design stage. By adopting this approach integrity is *baked-into* the design.

In summary, *ISO/IEC Standard 17799: Information technology—Security Techniques—Code of Practice for Information Security Management* (ISO/IEC 2000) states that integrity is "one of the three central pillars of securing corporate information assets along with confidentiality and availability" (Flowerday and von

Table 12.9 Design categories, design principles and life cycle stages

Design category	Design principle	C	D&C	T&E	O&M	R&D
Security Foundation	1. Establish a sound security policy as the "foundation" for design	☑	☑	☑	☑	☑
	2. Treat security as an integral part of the overall system design	☑	☑	☑	☑	☑
	3. Clearly delineate the physical and logical security boundaries governed by associated security policies	☑	☑	☑	☑	
	4. Ensure that developers are trained in how to develop secure software	☑	☑	☑		
Risk based	5. Reduce risk to an acceptable level	☑	☑	☑	☑	☑
	6. Assume that external systems are insecure	☑	☑	☑	☑	☑
	7. Identify potential trade-offs between reducing risk and increased costs and decrease in other aspects of operational effectiveness	☑	☑		☑	
	8. Implement tailored system security measures to meet organizational security goals	☑	☑	☑	☑	☑
	9. Protect information while being processed, in transit, and in storage	☑	☑	☑	☑	☑
	10. Consider custom products to achieve adequate security	☑	☑	☑	☑	
	11. Protect against all likely classes of "attacks"	☑	☑	☑	☑	☑
Ease of use	12. Where possible, base security on open standards for portability and interoperability	☑	☑	☑		
	13. Use common language in developing security requirements	☑	☑		☑	
	14. Design security to allow for regular adoption of new technology, including a secure and logical technology upgrade process		☑	☑	☑	
	15. Strive for operational ease of use	☑	☑	☑	☑	

(continued)

Table 12.9 (continued)

Design category	Design principle	C	D&C	T&E	O&M	R&D
Increase resilience	16. Implement layered security (ensure no single point of vulnerability)	☑	☑	☑	☑	☑
	17. Design and operate an IT system to limit damage and to be resilient in response	☑	☑		☑	
	18. Provide assurance that the system is, and continues to be, resilient in the face of expected threats	☑	☑	☑	☑	☑
	19. Limit or contain vulnerabilities		☑	☑	☑	
	20. Isolate public access systems from mission critical resources (e.g., data, processes, etc.)	☑	☑	☑	☑	
	21. Use boundary mechanisms to separate computing systems and network infrastructures		☑	☑	☑	
	22. Design and implement audit mechanisms to detect unauthorized use and to support incident investigations	☑	☑	☑	☑	
	23. Develop and exercise contingency or disaster recovery procedures to ensure appropriate availability	☑	☑	☑	☑	
Reduce vulnerabilities	24. Strive for simplicity	☑	☑	☑	☑	
	25. Minimize the system elements to be trusted	☑	☑	☑	☑	
	26. Implement least privilege	☑	☑	☑	☑	
	27. Do not implement unnecessary security mechanisms	☑	☑	☑	☑	☑
	28. Ensure proper security in the shutdown or disposal of a system		☑		☑	☑
	29. Identify and prevent common errors and vulnerabilities		☑	☑	☑	

(continued)

Table 12.9 (continued)

Design category	Design principle	C	D&C	T&E	O&M	R&D
Design with network in mind	30. Implement security through a combination of measures distributed physically and logically	☑	☑	☑	☑	☑
	31. Formulate security measures to address multiple overlapping information domains	☑	☑	☑	☑	
	32. Authenticate users and processes to ensure appropriate access control decisions both within and across domains	☑	☑	☑	☑	
	33. Use unique identities to ensure accountability	☑	☑	☑	☑	

Solms 2005, p. 606). Integrity is particularly concerned with the preserving the confidentiality and integrity of the information contained within the system. The next section will discuss how the non-functional requirements for accuracy, correctness, efficiency, and integrity may be measured and evaluated.

12.6 A Method for Evaluating Accuracy, Correctness, Efficiency, and Integrity

The ability to understand, measure, and evaluate the non-functional requirements for accuracy, correctness, efficiency, and integrity when included as requirements in a system is an important element of systems design. Having the ability to measure and evaluate each of these non-functional requirements provides additional perspectives and insight into the future performance and viability of all elements of the system being designed.

With a basic understanding of accuracy, correctness, efficiency, and integrity and how they are used in systems design endeavors, measurement may be addressed. As stated in other chapters, measurement is challenging because each of these non-functional requirements are subjective, qualitative measures which differ greatly from objective, quantitative measures. How to approach a subjective, qualitative measure will be reviewed.

12.6.1 Development of Measurement Scales

As with the other qualitative non-functional requirements, in order to satisfactorily evaluate accuracy, correctness, efficiency, and integrity, both the presence (yes or no) and quality of the effort (how well) to provide each of the non-functional requirements during system design endeavors must be addressed. To support this goal objects with a specific measureable attribute will be developed. The establishment of the measures is important because they are the link between what is observed in the real-world, and as such represent the empirical facts about the system and the construct for accuracy, correctness, efficiency, and integrity devised as evaluation points.

12.6.1.1 Scales for Accuracy, Correctness, Efficiency, and Integrity

As discussed during the development of the scales for in Chaps. 7 through 11, the selection of a measurement scale is an important element. Because the non-functional requirements for accuracy, correctness, efficiency, and integrity have no natural origin or empirically defined distance, an ordinal scale is an appropriate scale for measuring these criteria. As discussed in Chaps. 7 through 11, a five-point Likert scale will be invoked (Lissitz and Green 1975) in order to ensure improved reliability.

12.6.1.2 Proposed Scales

Scales are characterized as either a *proposed* scale or a scale. "A proposed scale is one that some investigator(s) put forward as having the requisite properties, and if it is indeed shown to have them, then it is recognized as a scale" (Cliff 1993, p. 65). As in the previous chapters, the use of the word scale is referring to *proposed scales*.

12.6.1.3 Proposed Measurement Scale for Accuracy, Correctness, Efficiency, and Integrity

Armed with a construct, measurement attributes, and an appropriate scale type, the measures for accuracy, correctness, efficiency, and integrity are constructed. In order to evaluate these, two essential questions must be answered. One addresses the presence (yes or no) and the other addresses the quality of the effort (how well) to provide effective and meaningful levels of accuracy, correctness, efficiency, and integrity during the system's design. Each of the four criteria are measurement constructs and have a specific question, shown in Table 12.10, which may be used to evaluate each one's contribution to a system's other viability concerns.

The answer to each question in Table 12.10 will be scored using the five-point Likert measures in Table 12.11.

The summation of the four constructs in Eq. 12.4 will be the measure of the degree of *other viability* in a system design endeavor.

Table 12.10 Measurement questions for other viability concerns

Measurement construct	Adaptation concern for measurement
$V_{accuracy}$	Does the system include a formal relationship for performance characteristics that is represented by accuracy, precision, and trueness?
$V_{correctness}$	Does the system include formal verification and validation activities that ensure that the system is correctly designed and operating through comparison with the system's specified requirements?
$V_{efficiency}$	Does the system evaluate efficiency using a calculation that is a ratio of the actual resources required to produce results against the benchmarked standard?
$V_{integrity}$	Does the system evaluate integrity by measuring its ability to ensure program correctness, noninterference, and information assurance?

Table 12.11 Other viability measurement question Likert scale

Measure	Descriptor	Measurement criteria
0.0	None	No objective quality evidence is present
0.5	Limited	Limited objective quality evidence is present
1.0	Nominal	Nominal objective quality evidence is present
1.5	Wide	Wide objective quality evidence is present
2.0	Extensive	Extensive objective quality evidence is present

Table 12.12 Four-level structural map for measuring other viability concerns

Level	Role
Concern	Systems viability
Attribute	Other viability concerns
Metrics	Accuracy, correctness, efficiency, and integrity
Measurable characteristics	Sum of (1) accuracy ($V_{accuracy}$), (2) correctness ($V_{correctness}$), (3) efficiency ($V_{efficiency}$), and (4) integrity ($V_{integrity}$)

Expanded Equation for Other Viability Concerns

$$V_{other} = V_{accuracy} + V_{correctness} + V_{efficiency} + V_{integrity} \tag{12.4}$$

12.6.2 Measuring Accuracy, Correctness, Efficiency, and Integrity

In each of the previous chapters the importance of being able to measure each non-functional attribute was stressed. A structural mapping that relates core viability concerns to four specific metrics and measurement entities is required. The four-level construct for other viability concerns is presented in Table 12.12.

12.7 Summary

In this chapter, the non-core or *other viability concerns* have been addressed. These include four non-functional requirements: (1) accuracy, (2) correctness, (3) efficiency, and (4) integrity. In each case, a formal definition has been provided along with additional explanatory definitions and terms. The ability to effect each of the four the non-functional requirements during the design process has also been addressed. Finally, a formal metric and measurement characteristic have been proposed for evaluating other viability concerns through metrics for accuracy, correctness, efficiency, and integrity.

The next chapter will discuss the use of the complete taxonomy of non-functional requirements as part of the purposeful design of complex systems during systems design endeavors.

References

Ackoff, R. L. (1999). *Ackoff's best: His classic writings on management*. New York: Wiley.

Adams, K. M. (2011). Systems principles: foundation for the SoSE methodology. *International Journal of System of Systems Engineering, 2*(2/3), 120–155.

Adams, K. M., Hester, P. T., Bradley, J. M., Meyers, T. J., & Keating, C. B. (2014). Systems theory: The foundation for understanding systems. *Systems Engineering, 17*(1), 112–123.

ANSI/EIA. (1998). *ANSI/EIA standard 632: Processes for engineering a system*. Arlington, VA: Electronic Industries Alliance.

Biba, M. J. (1975). *Integrity considerations for secure computer systems (MTR 3153)*. Bedford, MA: MITRE.

Blundell, J. K., Hines, M. L., & Stach, J. (1997). The measurement of software design quality. *Annals of Software Engineering, 4*(1), 235–255.

Boehm, B. W., Brown, J. R., & Lipow, M. (1976). Quantitative evaluation of software quality. In R. T. Yeh & C. V. Ramamoorthy (Eds.), *Proceedings of the 2nd international conference on software engineering* (pp. 592–605). Los Alamitos, CA: IEEE Computer Society Press.

Boulding, K. E. (1966). *The impact of social sciences*. New Brunswick, NJ: Rutgers University Press.

Bowen, J. P., & Hinchey, M. G. (1998). *High-integrity system specification and design*. London: Springer.

Bowen, T. P., Wigle, G. B., & Tsai, J. T. (1985). *Specification of software quality attributes: Software quality evaluation guidebook (RADC-TR-85-37)* (Vol. III). Griffiss Air Force Base, NY: Rome Air Development Center.

Bresciani-Turroni, C. (1937). On Pareto's law. *Journal of the Royal Statistical Society, 100*(3), 421–432.

Cavano, J. P., & McCall, J. A. (1978). A framework for the measurement of software quality. *SIGSOFT Software Engineering Notes, 3*(5), 133–139.

Cherns, A. (1976). The principles of sociotechnical design. *Human Relations, 29*(8), 783–792.

Cherns, A. (1987). The principles of sociotechnical design revisited. *Human Relations, 40*(3), 153–161.

Churchman, C. W., & Ratoosh, P. (Eds.). (1959). *Measurement: Definitions and theories*. New York: Wiley.

Cliff, N. (1993). What is and isn't measurement. In G. Keren & C. Lewis (Eds.), *A Handbook for Data Analysis in the Behavioral Sciences: Methodological Issues* (pp. 59–93). Hillsdale, NJ: Lawrence Erlbaum Associates.

Colbert, B. A. (2004). The complex resource-based view: Implications for theory and practice in strategic human resource management. *Academy of Management Review, 29*(3), 341–358.

Courtney, R. H., & Ware, W. H. (1994). What do we mean by integrity? *Computers & Security, 13* (3), 206–208.

Creedy, J. (1977). Pareto and the distribution of income. *Review of Income and Wealth, 23*(4), 405–411.

De Bièvre, P. (2006). Accuracy versus uncertainty. *Accreditation and Quality Assurance, 10*(12), 645–646.

Del Mar, D. (1985). *Operations and industrial management*. New York: McGraw-Hill.

Dodge, Y. (2008). *The concise Encyclopedia of statistics*. New York: Springer.
Drucker, P. F. (2001). *The essential Drucker: The best of 60 years of Peter Drucker's essential writings on management*. New York: Harper Collins Publishers.
EIA. (1999). *Electronic Industries Association (EIA) Interim Standard (IS) 731: The systems engineering capability model*. Arlington, VA: Electronic Industries Association.
Flowerday, S., & von Solms, R. (2005). Real-time information integrity = system integrity + data integrity + continuous assurances. *Computers & Security, 24*(8), 604–613.
Foley, S. N. (2003). A nonfunctional approach to system integrity. *IEEE Journal on Selected Areas in Communications, 21*(1), 36–43.
Georg, G., France, R., & Ray, I. (2003). Designing high integrity systems using aspects. In M. Gertz (Ed.), *Integrity and internal control in information systems V* (Vol. 124, pp. 37–57). New York: Springer.
Hahn, G. J., Hill, W. J., Hoerl, R. W., & Zinkgraf, S. A. (1999). The impact of six sigma improvement—a glimpse into the future of statistics. *The American Statistician, 53*(3), 208–215.
IEEE. (2005). *IEEE Standard 1220: Systems engineering—application and management of the systems engineering process*. New York: Institute of Electrical and Electronics Engineers.
IEEE. (2012). *IEEE Standard 1012: IEEE standard for system and software verification and validation*. New York: The Institute of Electrical and Electronics Engineers.
IEEE, & ISO/IEC. (2008). IEEE and ISO/IEC Standard 12207: Systems and software engineering —software life cycle processes. New York and Geneva: Institute of Electrical and Electronics Engineers and the International Organization for Standardization and the International Electrotechnical Commission.
IEEE, & ISO/IEC. (2010). IEEE and ISO/IEC Standard 24765: Systems and software engineering —vocabulary. New York and Geneva: Institute of Electrical and Electronics Engineers and the International Organization for Standardization and the International Electrotechnical Commission.
ISO/IEC. (2000). ISO/IEC Standard 17799: Information technology—security techniques—code of practice for information security management. Geneva: International Organization for Standardization and the International Electrotechnical Commission.
Jacob, J. (1991). The basic integrity theorem. In *Proceedings of the Computer Security Foundations Workshop IV* (pp. 89–97). Los Alamitos, CA: IEEE Computer Society Press.
JCGM. (2012). JCGM Standard 200: International vocabulary of metrology—basic and general concepts and associated terms (VIM) Sèvres, France: Joint Committee for Guides in Metrology, International Bureau of Weights and Measures (BIPM).
Li, P., Mao, Y., & Zdancewic, S. (2003). Information integrity policies. In *Proceedings of the 1st International Workshop on Formal Aspects in Security and Trust (FAST'03)* (pp. 39–51). Pisa, Italy.
Lissitz, R. W., & Green, S. B. (1975). Effect of the number of scale points on reliability: A Monte Carlo approach. *Journal of Applied Psychology, 60*(1), 10–13.
Madni, A. M. (2012). Elegant systems design: Creative fusion of simplicity and power. *Systems Engineering, 15*(3), 347–354.
McCall, J. A., & Matsumoto, M. T. (1980). *Software quality measurement manual (RADC-TR-80-109-Vol-2)*. Griffiss Air Force Base, NY: Rome Air Development Center.
Menditto, A., Patriarca, M., & Magnusson, B. (2007). Understanding the meaning of accuracy, trueness and precision. *Accreditation and Quality Assurance, 12*(1), 45–47.
Miller, G. A. (1956). The magical number seven, plus or minus two: Some limits on our capability for processing information. *Psychological Review, 63*(2), 81–97.
Minnich, I. (2002). EIA IS 731 compared to CMMISM-SE/SW. *Systems Engineering, 5*(1), 62–72.
Murphy, R. B. (1969). On the meaning of precision and accuracy. In H. H. Ku (Ed.), *Precision measurement and calibration: Selected NBS papers on statistical concepts and procedures (NBS special publication 300)* (pp. 357–360). Washington, DC: Government Printing Office.
Rabinovich, S. G. (2013). *Evaluating measurement accuracy*. New York: Springer.
Sabelfeld, A., & Myers, A. C. (2003). Language-based information-flow security. *IEEE Journal on Selected Areas in Communications, 21*(1), 5–19.

Sanders, R. E. (1987). The Pareto principle: Its use and abuse. *The Journal of Services Marketing, 1*(2), 37–40.

Sandhu, R. S., & Jajodia, S. (1993). Data and database security and controls. In H. F. Tipton & Z. G. Ruthbert (Eds.), *Handbook of information security management* (pp. 481–499). Boston: Auerbach.

Sandhu, R. S., & Jajodia, S. (1994). Integrity mechanisms in database management systems. In M. D. Abrams, S. Jajodia, & H. J. Podell (Eds.), *Information security: An integrated collection of essays* (pp. 617–635). Los Alamitos, CA: IEEE Computer Society Press.

Simon, H. A. (1974). How big is a chunk? *Science, 183*(4124), 482–488.

Sproles, N. (2001). The difficult problem of establishing measures of effectiveness for command and control: A systems engineering perspective. *Systems Engineering, 4*(2), 145–155.

Stoneburner, G., Hayden, C., & Feringa, A. (2004). *Engineering principles for information technology security (A baseline for achieving security), [NIST special publication 800-27 Rev A]*. Gaithersburg, MD: National Institute of Standards and Technology.

Szilagyi, A. D. (1984). *Management and performance* (2nd ed.). Glenview, IL: Scotts, Foresman and Company.

Thuesen, G. J., & Fabrycky, W. J. (1989). *Engineering economy*. Englewood Cliffs, NJ: Prentice-Hall.

Warfield, J. N. (1999). Twenty laws of complexity: Science applicable in organizations. *Systems Research and Behavioral Science, 16*(1), 3–40.

Wells, C., Ibrahim, L., & LaBruyere, L. (2003). A new approach to generic attributes. *Systems Engineering, 6*(4), 301–308.

Part VI
Conclusion

Chapter 13
Conclusion

Abstract The design of systems and components is a crucial element that affects both the cost and efficacy of products produced for the world economy. Design is a characteristic function of engineering. The structure of engineering education underwent a major shift after WWII. The nationwide shift toward a more science-based curricula for all levels of education led to design type courses being devalued and even omitted in engineering education. Recent efforts to re-invigorate design in both undergraduate and graduate engineering programs in the United States have re-emphasized the role of design in the engineering curricula. The current text has been developed to address a unique topic in engineering design—non-functional requirements in systems analysis and design endeavors, thereby seeking to fill a perceived void in the existing engineering literature.

13.1 Position and Importance of Engineering Design

The competitive difficulties in the world market that are faced by products manufactured in the United States has been attributed to a variety of causes. The MIT Commission on Industrial Productivity addressed the recurring weaknesses of American industry that continue to threaten the country's standard of living and its position in the world economy (Dertouzos et al. 1989). "To regain world manufacturing leadership, we need to take a more strategic approach by also improving our engineering design practices" (Dixon and Duffey 1990, p. 9).

"Market loss by U.S. companies is due to design deficiencies more than manufacturing deficiencies" (Dixon and Duffey 1990, p. 13). The importance of engineering design and its associated activities, especially when compared to more glamorous activities such as marketing and sales, directly affects the cost and long-term efficacy of products produced for the world market.

> Engineering design is a crucial component of the industrial product realization process. It is estimated that 70 % or more of the life cycle cost of a product is determined during design. (NRC 1991, p. 1)

K.M. Adams, *Non-functional Requirements in Systems Analysis and Design*,
Topics in Safety, Risk, Reliability and Quality 28,
DOI 10.1007/978-3-319-18344-2_13

Table 13.1 Primary activities of engineers in 1982 [adapted from Table 3 in (NRC 1985, p. 91)]

Activity	Percentage
Research	4.7
Development, including design	27.9
R&D management	8.7
Other management	19.3
Teaching	2.1
Production/inspection	16.6
Unreported	20.7

> According to General Motors executives, 70 % of the cost of manufacturing truck transmissions is determined in the design stage. (Whitney 1988, p. 83)
> A study at Rolls-Royce revealed that design activities account for 80 % (i.e., design schemes 50 % and detail drawings 30 %) of the final production costs of 2000 components. (Corbett and Crookall 1986)

Clearly, design activities and the engineers tasked with implementing them are important elements of the global economy. Design is a characteristic function within the field of engineering. Although not all engineers are directly involved in design, a 1982 study of the primary activities of engineers displayed in Table 13.1, reported that 28 % were working in development and design related activities.

13.2 Education in Engineering Design

Nobel Laureate Herbert A. Simon [1916–2001] recounted how major universities and colleges, after World War II, shifted their curriculums to reflect their movement toward the natural sciences and away from the *sciences of the artificial*. Simon (1996) uses the term *sciences of the artificial* to refer to the tasks associated with how to "teach about artificial things: how to make artifacts that have desired properties and how to design" (p. 111). There were deleterious effects as a result of the shift to a science-based foundation.

- *It is now widely believed that U.S. industry's extended period of world dominance in product design, manufacturing innovation, process engineering, productivity, and market share has ended* (NRC 1986, p. 5).
- *These changes include restructuring to emphasize the engineering sciences as a coherent body of knowledge, the introduction of new disciplines, the creation of an extensive system of research and graduate programs, and the partial integration of computers into curricula. While these improvements were taking place, the state of engineering design education was steadily deteriorating with the result that today's engineering graduates are poorly equipped to utilize their scientific, mathematical, and analytical knowledge in the design of components, processes, and systems* (NRC 1991, p. 35).
- *Engineering schools gradually became schools of physics and mathematics* (Simon 1996, p. 111).

- *An inevitable by-product of the science revolution was that engineering design, because it did not have a formalized, quantitative, teachable core body of knowledge, was largely eliminated from engineering curricula. Instead, engineers were expected to learn design on the job. Indeed, the development of a formalized approach to engineering design remains an open challenge to the engineering professoriate* (Tadmor 2006, p. 34).

As a result, most engineering curricula have insufficient levels of instruction and practical applications in design. The deficiencies exist at both the undergraduate and graduate levels. At the undergraduate level of engineering, design education weaknesses include "weak requirements for design content in engineering curricula (many institutions do not meet even existing accreditation criteria); lack of truly interdisciplinary teams in design courses; and fragmented, discipline-specific, and uncoordinated teaching" (NRC 1991, p. 2). At the graduate level, "there are simply too few strong graduate programs focusing on modern design methodologies and research to produce the qualified graduates needed by both industry and academe" (NRC 1991, p. 3). In addition, the deficiencies present in the undergraduate curricula directly affect the ability to teach design at the graduate level. This is because the inadequate undergraduate design experience ill prepares students to take graduate design courses and forces graduate courses to become remedial in nature. Finally, the monetary incentives are forcing graduate programs in engineering to admit students with non-engineering degrees. These students, like the ill-prepared undergraduate engineers, have no formal training in design, adding additional pressure to structure graduate courses in design to address topics at the remedial level.

In response, *The Committee on Engineering Design Theory and Methodology* of the National Research Council, recommended the following actions:

- Undergraduate engineering design education must (NRC 1991, p. 36):

    1. Show how the fundamental engineering science background is relevant to effective design;
    2. Teach students what the design process entails and familiarize them with the basic tools of the process;
    3. Demonstrate that design involves not just function but also producibility, cost, customer preference, and a variety of life cycle issues; and
    4. Convey the importance of other subjects such as mathematics, economics, and manufacturing.

- Graduate design education should be directed toward (NRC 1991, p. 37):

    1. Developing competence in advanced design theory and methodology;
    2. Familiarizing graduate students with state-of-the-art ideas in design, both from academic research and from worldwide industrial experience and research;
    3. Providing students with working experience in design;

4. Immersing students in the entire spectrum of design considerations, preferably during industrial internships; and
5. Having students perform research in engineering design.

Finally, it is important to note that the Accreditation Board for Engineering and Technology (ABET), the organization responsible for the accreditation of all engineering programs in the United States, requires all undergraduate engineering curricula, in its general requirements for baccalaureate level programs to satisfy three criterion associated with design (ABET 2013):

1. *Criterion 3—Student outcomes*: (c) an ability to design a system, component, or process to meet desired needs within realistic constraints such as economic, environmental, social, political, ethical, health and safety, manufacturability, and sustainability; and (d) an ability to function on multidisciplinary teams.
2. *Criterion 5—Curriculum*: (b) one and one-half years of engineering topics, consisting of engineering sciences and engineering design appropriate to the student's field of study. The engineering sciences have their roots in mathematics and basic sciences but carry knowledge further toward creative application. These studies provide a bridge between mathematics and basic sciences on the one hand and engineering practice on the other. Engineering design is the process of devising a system, component, or process to meet desired needs. It is a decision-making process (often iterative), in which the basic sciences, mathematics, and the engineering sciences are applied to convert resources optimally to meet these stated needs. Note: One year is the lesser of 32 semester hours (or equivalent) or one-fourth of the total credits required for graduation.
3. *Criterion 5—Curriculum*: Students must be prepared for engineering practice through a curriculum culminating in a major design experience based on the knowledge and skills acquired in earlier course work and incorporating appropriate engineering standards and multiple realistic constraints.

The ABET expectation is that design topics are taught throughout the undergraduate curricula. Specific design related tasks are addressed in the associated engineering topics and that, when combined, these courses will contain all the design information necessary to complete the major design experience (e.g., a capstone course or senior design project). Furthermore, the major design experience focused on a design and will not introduce new topics or materials about the design process. Despite the ABET requirements and expectations, the capstone design course or senior design project often lacks focus as the culminating engineering experience.

> Often, too much is expected of these senior design courses when prior courses have failed to provide sound preparation for them. When, for example, a senior design course is a student's only exposure to integrated design activities such as concurrent design, detailed consideration of alternatives and constraints, significant economic analyses, and working as part of a team, the experience is likely to be shallow. (NRC 1991, p. 40)

13.3 Position of This Text Within Engineering Design

This text is positioned as a guide for a course in engineering design that focuses on the elements of the design that do not provide a direct function in support of the stakeholder's processes. The non-functional requirements addressed in the text are elements of the design that affect performance of the entire system, and are not attributable to any one specific function or process mandated by the system's stakeholders. In fact, the system's stakeholders may not recognize terms such as survivability, robustness, and self-descriptiveness. It is the job of the engineer conducting the design to ensure that appropriate system-wide, non-functional requirements are addressed and invoked in order to effectively treat sustainment, design, adaptation and viability concerns.

As such, this text is satisfying a subset of the goals established for engineering design that were addressed in the previous section. Specifically:

- Undergraduate engineering design education actions to:

  1. Teach students what the design process entails and familiarize them with the basic tools of the process;
  2. Demonstrate that design involves not just function but also producibility, cost, customer preference, and a variety of life cycle issues; and

- Graduate design education actions to:

  3. Developing competence in advanced design theory and methodology;
  4. Familiarizing graduate students with state-of-the-art ideas in design, both from academic research and from worldwide industrial experience and research;

- ABET Accreditation Criteria:

  5. *Criterion 3—Student outcomes*: (c) an ability to design a system, component, or process to meet desired needs within realistic constraints such as economic, environmental, social, political, ethical, health and safety, manufacturability, and sustainability; and (d) an ability to function on multidisciplinary teams.
  6. *Criterion 5—Curriculum*: (b) Engineering design is the process of devising a system, component, or process to meet desired needs. It is a decision-making process (often iterative), in which the basic sciences, mathematics, and the engineering sciences are applied to convert resources optimally to meet these stated needs.

13.4 Summary

Undergraduate and graduate education in engineering is the foundation upon which all engineering graduates base their professional activities. As a result, the academic community, and in particular those who use their own training and education to lead courses of instruction in design must provide clearly articulated scholarly materials that directly support these courses of instruction. This text has been developed to address a unique topic in engineering design. To date, non-functional requirements have only been addressed within highly focused sub-disciplines of engineering (i.e., reliability, maintainability, availability; traceability; testability; survivability; etc.). The wider engineering community has not had access to materials that permit them to develop a holistic, systemic perspective for non-functional requirements that regularly affect the entire system. Generally, engineers participate on transdisciplinary teams where experts in this or that non-functional requirement (e.g., the dreaded ilities) come in and assess the system's ability to satisfy the particular non-functional requirement. When the experts produce the list of deficiencies they leave and the component and sub-system level engineers scratch their heads and are faced with changing the design to ensure compliance with this or that non-functional requirement. Having a basic understanding of how the principal non-functional requirements affect the sustainment, design, adaptability, and viability concerns of a system, at a high-level, will fill help a void in the existing engineering literature.

References

ABET. (2013). *Criteria for accrediting engineering programs: Effective for reviews during the 2014–2015 accreditation cycle (E001 of 24 Feb 2014)*. Baltimore, MD: Accreditation Board for Engineering and Technology.

Corbett, J., & Crookall, J. R. (1986). Design for economic manufacture. *CIRP Annals—Manufacturing Technology, 35*(1), 93–97.

Dertouzos, M. L., Solow, R. M., & Lester, R. K. (1989). *Made in America: Regaining the productive edge*. Cambridge, MA: MIT Press.

Dixon, J. R., & Duffey, M. R. (1990). The neglect of engineering design. *California Management Review, 32*(2), 9–23.

NRC. (1985). *Engineering education and practice in the United States: Foundations of our techno-economic future*. Washington, DC: National Academies Press.

NRC. (1986). *Toward a new era in U.S. manufacturing: The need for a national vision* Washington, DC: National Academies Press.

NRC. (1991). *Improving engineering design: Designing for competitive advantage*. Washington, DC: National Academy Press.

Simon, H. A. (1996). *The sciences of the artificial* (3rd ed.). Cambridge, MA: MIT Press.

Tadmor, Z. (2006). Redefining engineering disciplines for the twenty-first century. *The Bridge, 36* (2), 33–37.

Whitney, D. E. (1988). Manufacturing by design. *Harvard Business Review, 66*(4), 83–91.

Index

K.M. Adams, *Non-functional Requirements in Systems Analysis and Design*, Topics in Safety, Risk, Reliability and Quality 28,
DOI 10.1007/978-3-319-18344-2

GPSR Compliance
The European Union's (EU) General Product Safety Regulation (GPSR) is a set of rules that requires consumer products to be safe and our obligations to ensure this.

If you have any concerns about our products, you can contact us on

ProductSafety@springernature.com

In case Publisher is established outside the EU, the EU authorized representative is:

Springer Nature Customer Service Center GmbH
Europaplatz 3
69115 Heidelberg, Germany

www.ingramcontent.com/pod-product-compliance
Lightning Source LLC
Chambersburg PA
CBHW072145090226
39397CB00004B/273
* 9 7 8 3 3 1 9 1 8 3 4 3 5 *